Lecture Notes in Computer Science 8847

Commenced Publication in 1973
Founding and Former Series Editors:
Gerhard Goos, Juris Hartmanis, and Jan van Leeuwen

More information about this series at http://www.springer.com/series/7411

Jie Gao · Alon Efrat
Sándor P. Fekete · Yanyong Zhang (Eds.)

Algorithms for Sensor Systems

10th International Symposium on Algorithms
and Experiments for Sensor Systems,
Wireless Networks and Distributed Robotics,
ALGOSENSORS 2014
Wroclaw, Poland, September 12, 2014
Revised Selected Papers

 Springer

Editors
Jie Gao
Stony Brook University
Stonybrook, NY
USA

Alon Efrat
University of Arizona
Tucson, AZ
USA

Sándor P. Fekete
TU Braunschweig
Braunschweig, Niedersachsen
Germany

Yanyong Zhang
Rutgers University
New Brunswick, NJ
USA

ISSN 0302-9743   ISSN 1611-3349   (electronic)
Lecture Notes in Computer Science
ISBN 978-3-662-46017-7   ISBN 978-3-662-46018-4   (eBook)
DOI 10.1007/978-3-662-46018-4

Library of Congress Control Number: 2014958662

LNCS Sublibrary: SL5 – Computer Communication Networks and Telecommunications

Springer Heidelberg New York Dordrecht London

Printed on acid-free paper

Springer-Verlag GmbH Berlin Heidelberg is part of Springer Science+Business Media
(www.springer.com)

Preface

ALGOSENSORS, the International Symposium on Algorithms and Experiments for Sensor Systems, Wireless Networks, and Distributed Robotics, is an international forum dedicated to the algorithmic aspects of wireless networks, static or mobile. The 10th edition of ALGOSENSORS was held on September 12 in Wroclaw, Poland, within the ALGO annual event.

Originally focused solely on sensor networks, ALGOSENSORS now covers more broadly algorithmic issues arising in all wireless networks of computational entities, including sensor networks, sensor-actuator networks, and systems of autonomous mobile robots. In particular, it focuses on the design and analysis of discrete and distributed algorithms, on models of computation and complexity, on experimental analysis, in the context of wireless networks, sensor networks, and robotic networks and on all foundational and algorithmic aspects of the research in these areas.

This year papers were solicited into three tracks: Sensor Network Algorithms (Track A), Wireless Networks and Distributed Robotics (Track B), and Experimental Algorithms (Track C).

In response to the call for papers, 20 submissions were received overall, out of which 10 papers were accepted after a rigorous reviewing process by the (joint) Program Committee, which involved at least three reviewers per paper. The committee had an online discussion and the final accepted list was agreed by all members of the committee. In addition to the technical papers, the program included an invited keynote talk by Dr. Phillip Gibbons (Intel Labs Pittsburgh). This volume contains the technical papers as well as a summary of the keynote talk. We would like to thank the Program Committee members, as well as the external reviewers, for their fundamental contribution in selecting the best papers resulting in a strong program. We would also like to warmly thank the ALGO/ESA 2014 organizers for kindly accepting the proposal of the Steering Committee to co-locate ALGOSENSORS with some of the leading events on algorithms in Europe.

October 2014

Jie Gao
Alon Efrat
Sándor P. Fekete
Yanyong Zhang

Organization

Technical Program Committee

Jie Gao (General Chair)	Stony Brook University, USA
Alon Efrat (Track Co-chair, Sensor Networks Algorithms)	University of Arizona, USA
Sándor P. Fekete (Track Co-chair, Wireless Networks and Distributed Robotics)	TU Braunschweig, Germany
Yanyong Zhang (Track Co-chair, Experiments)	Rutgers University, USA

Algorithms Track

Chris Gniady	University of Arizona, USA
Guy Grebla	Columbia University, USA
Loukas Lazos	University of Arizona, USA
Miao Jin	University of Louisiana, USA
Thienne Johnson	University of Arizona, USA
Valentin Polishchuk	Linköping University, Sweden
Andrea Richa	Arizona State University, USA
Rik Sarkar	University of Edinburgh, UK
Michael Segal	Ben-Gurion University of the Negev, Israel
Guoliang Xue	Arizona State University, USA

Robotics Track

Aaron Becker	Harvard University, USA
Thomas Erlebach	University of Leicester, UK
Michael Hemmer	TU Braunschweig, Germany
Alejandro López-Ortiz	University of Waterloo, Canada
James McLurkin	Rice University, USA
Friedhelm Meyer auf der Heide	University of Paderborn, Germany
Joe Mitchell	Stony Brook University, USA
Nicola Santoro	Carleton University, Canada
Christiane Schmidt	TU Braunschweig, Germany
Jukka Suomela	Aalto University, Finland

Experiments Track

Marios Angelopoulos	University of Geneva, Switzerland
James Gross	KTH, Sweden
Yuan He	Tsinghua University, China
Tommaso Melodia	University of Buffalo, USA
Hui Pan	Hong Kong University of Science and Technology, Hong Kong
Dario Pompili	Rutgers University, USA
Aaron Striegel	University of Notre Dame, USA
Niki Trigoni	University of Oxford, UK
Guiling Wang	New Jersey Institute of Technology, USA
Kai Xing	University of Science and Technology of China, China
Rong Zheng	McMaster University, Canada
Marco Zuniga	TU Delft, The Netherlands

Steering Committee

Josep Diaz	Universitat Politècnica de Catalunya, Spain
Magnus M. Halldorsson	Reykjavik University, Iceland
Bhaskar Krishnamachari	University of Southern California, USA
P.R. Kumar	Texas A&M University, USA
Sotiris Nikoletseas	University of Patras and CTI, Greece (Chair)
Jose Rolim	University of Geneva, Switzerland
Paul Spirakis	University of Patras and CTI, Greece
Adam Wolisz	TU Berlin, Germany

Conference Organization

Web Chair

Rik Sarkar	University of Edinburgh, UK

Publicity Co-chair

Marios Angelopoulos	University of Geneva, Switzerland
Chenren Xu	Rutgers University, USA

Algorithmic Challenges in M2M
(Invited Talk)

Phillip B. Gibbons

Intel Science and Technology Center for Cloud Computing,
Carnegie Mellon University,
Pittsburgh, PA, USA
`phillip.b.gibbons@intel.com`

Abstract. The Internet of Things promises a world of billions to trillions of smart objects/ devices, communicating machine-to-machine (M2M) and providing us valuable information and services. This talk highlights our recent work addressing several key algorithmic challenges that arise in this setting. Specifically, we focus on problems arising in aggregation, similarity search, and machine learning on M2M's massively distributed network. After surveying these results, we present in greater detail upper and lower bounds demonstrating the cost of fault tolerance in such networks. These bounds show that across a communication-time trade-off curve, aggregation algorithms that tolerate crash failures incur an exponential cost in communication relative to non-fault-tolerant algorithms.

Contents

Robot Planning

The Multi-source Beachcombers' Problem

Jurek Czyzowicz[1], Leszek Gąsieniec[2], Konstantinos Georgiou[3],
Evangelos Kranakis[4(✉)], and Fraser MacQuarrie[4]

[1] Department d'Informatique, Université du Québec en Outaouais,
Gatineau, QC, Canada
[2] Department of Computer Science, University of Liverpool, Liverpool, UK
[3] Department of Combinatorics and Optimization,
University of Waterloo, Waterloo, ON, Canada
[4] School of Computer Science, Carleton University,
Ottawa, ON, Canada
kranakis@scs.carleton.ca

Abstract. The Beachcombers' Problem (c.f. [1]) is an optimization
problem in which a line segment is to be searched by a set of mobile
robots, where each robot has a *searching speed* s_i and a *walking speed* w_i,
such that $s_i < w_i$. We explore a natural generalization of the Beachcombers' Problem, the t-Source Beachcombers' Problem (t-SBP), where
the robots are not constrained to start at a single source: Consider n
mobile robots labelled $1, 2, \ldots, n$. We choose t *sources* and we assign
each robot to one of them. The problem is to choose the sources and
develop *mobility schedules* (algorithms) for the robots which maximizes
the *speed* of the fleet, or minimize the time that the robots need to collectively search the domain. We propose several algorithms for solving
problems of this nature. We prove that 2-SBP is NP-hard even when
the robots have *identical walking* speeds. This contrasts with the case
when the robots have *identical search* speeds, where we give a polynomial time algorithm for t-SBP. We also give a 0.5569-approximation
for 2-SBP with arbitrary walking and searching speeds. For t-SBP with
arbitrary walking and searching speeds, we give an oblivious randomized
algorithm and prove that it provides an expected $1 - 1/e$ approximation,
asymptotically in t.

Keywords: Algorithm · Approximation · Mobile robots · Partitioning ·
Randomized · Schedule · Searching · Segment · Speed · Walking

1 Introduction

A continuous one-dimensional domain is to be explored collectively by n robots.
Each robot has *two* speeds: *walking* and *searching*. The first is the speed with
which it can *traverse* the domain, while the second is the speed with which it can

J. Czyzowicz and E. Kranakis—Research supported in part by NSERC Discovery
grant.

J. Gao et al. (Eds.): ALGOSENSORS 2014, LNCS 8847, pp. 3–21, 2015.
DOI: 10.1007/978-3-662-46018-4_1

perform a more *elaborate* task (like searching) on the domain. We use the analogy of the robots as *beachcombers* to emphasize that when searching a domain (e.g. a beach looking for items of value), robots move slower than if they were simply traversing the domain.

In the classical search problem, a searcher wishes to find an object of interest (target) located somewhere in a domain. The searcher executes a search by deploying a swarm of mobile agents, which are able to move (and search) in the domain with a single speed. By allowing agents the ability to traverse the domain at high speed (but not searching while doing so), the Beachcombers' Problem changes the nature of the question, since one needs to now consider the trade-off between walking and searching. There are many examples where *two speed* explorations are natural and useful. For example, *forensic search* would require that electronically stored information be searched more thoroughly, *code inspection* in programming may require more elaborate searching, as well as *foraging* and/or *harvesting* a field may take longer than walking. Similar scenarios could occur in search and rescue operations, allocating marketing, law enforcement, data structures, database applications, and artificial intelligence.

1.1   Preliminaries and Notation

The input to our problem will always contain a swarm of n robots. Each robot i has a *searching speed* s_i and a *walking speed* w_i, where $s_i < w_i$. A swarm is denoted by the tuple of speed vectors $(\mathbf{s}, \mathbf{w})$. A swarm where the walking speeds are all the same is called *W-uniform*. Similarly in an *S-uniform* swarm, all searching speeds are the same.

Robots are placed initially at a *source*, which determines the initial position and direction of movement for the robot. There can be two sources at any given point on the domain (considered in this paper to be an interval), one for each direction of movement. This definition is the reason the robots in 1-SBP must start at an endpoint: since there is only one source, if the source is an interior point of the interval, only points that lie on the interval in the direction of movement of the robots will ever be searched. At any moment a robot can be: *idle*, *searching*, or *walking*. When searching or walking, it is moving with speed no greater than s_i and w_i respectively. Robots can switch states instantaneously, as many times as needed, and at any time. We assume that the robots' schedules are controlled by a centralized scheduler, that the robots start at the same time, and that they can cross over each other during their operation.

We are interested in providing *mobility schedules* for solving the Multi-source Beachcombers' Problem. A *schedule* consists of (a) a partition of the swarm into t groups, (b) a choice of sources, one for each of the groups, and (c) a mode-schedule which specifies at each moment the state of any robot, i.e. idle, searching or walking. A schedule is called *feasible* if there is a *finishing time* T after which all points of the domain have been searched by at least one robot. Hereafter, all schedules referred to will be assumed to be feasible.

t-Source Beachcombers' Problem (t-SBP). Consider an interval $I_L = [0, L]$ and n robots $r_1, r_2, \ldots, r_n$, each robot r_i having searching speed s_i and walking speed w_i, such that $s_i < w_i$. The t-Source Beachcombers' Problem consists of finding a correct searching schedule $\mathcal{A}$ of I_L which has at most t unique sources, and which maximizes the *speed* with which the robots search the interval. The speed $S_{\mathcal{A}}$ of the solution to the Beachcombers' Problem is defined to be $S_{\mathcal{A}} = L/T$, where T is the finishing time of $\mathcal{A}$.[1]

Whenever it is clear from the context, we may silently assume the normalization $L = 1$. 1-SBP was studied in [1] where the following properties of optimal schedules were proven.

Lemma 1 (Czyzowicz et al. [1]). *In every optimal schedule of 1-SBP:*

- *at any moment, each robot moves in full search or walking speed, and it is never idle;*
- *all robots terminate their work simultaneously and each robot completes its work by searching some non-empty interval;*
- *each robot searches a continuous interval;*
- *for any two robots r_i, r_j with $w_i < w_j$, robot r_i searches a sub-interval closer to the starting point than the sub-interval of robot r_j.*

These properties were used to show the optimality of Algorithm 1, where the search intervals are calculated using the formula given in Lemma 2.

Algorithm 1. Comb (Czyzowicz et al. [1])

Require: swarm $(\mathbf{s}, \mathbf{w})$.
Ensure: Schedule of the swarm.
 1: Sort the robots in non-decreasing walking speeds.
 2: Calculate search intervals $c_1, \ldots, c_n$.
 3: All robots start processing the domain simultaneously. In particular,
 4: **for** $i = 1, \ldots, n$ (in parallel) **do**
 5:   Robot i first walks the interval length $\sum_{j=1}^{i-1} c_j$, and then searches interval length c_i.
 6: **end for**

Lemma 2 (Czyzowicz et al. [1]). *For the 1-SBP problem, let robots $r_1, \ldots, r_n$ be ordered in non-decreasing walking speed, and suppose that T_{opt} is the time of the optimal schedule. Then, the segment to be searched may be partitioned into successive sub-segments of lengths $c_1, c_2, \ldots, c_n$ and the optimal schedule assigns to robot r_i the ith interval of length c_i, where the length c_i satisfies the following recursive formula, and where we assume, without loss of generality, that $w_0 = 0$ and $w_1 = 1$.[2]*

$$c_0 = 0; \qquad c_k = \frac{s_k}{w_k}\left(\left(\frac{w_{k-1}}{s_{k-1}} - 1\right)c_{k-1} + T_{opt}(w_k - w_{k-1})\right), \quad k \geq 1 \qquad (1)$$

---

[1] Note that maximizing the speed of a feasible schedule is equivalent to minimizing its finishing time.

[2] We set $w_0 = 0$ and $w_1 = 1$ for notational convenience, so that (1) holds. Note that w_0 does not correspond to any robot, while w_1 is the walking speed of the robot that will search the first sub-interval, and so will never enter walking mode, hence, w_1 does not affect our solution.

The concept of *search power* was introduced in [1] to quantify the solution cost for 1-SBP. For a swarm of robots N, i.e. a collection N of robots associated with certain walking and searching speeds, it was shown that the solution cost for 1-SBP is $g(N)$, where $g(A)$ is defined below. More simply put, search power is the speed of the optimal schedule with 1 source, i.e. if N is the collection of robots with speeds $(\mathbf{s}, \mathbf{w})$, then the search power $g(N)$ is the optimal speed of the swarm for 1-SBP, and it is the result of the schedule of Algorithm 1.

Definition 1 (Search Power). *Consider a swarm with attributes* (s_i, w_i), *with* $s_i < w_i$, $i = 1, \ldots, n$. *We define the search power of any subset of the robots using a real function* $g : 2^{[n]} \mapsto \mathbb{R}^+$ *as follows: For any subset* A, *first sort the items in non-decreasing weights* w_i, *and let* $w_1^A, \ldots, w_{|A|}^A$ *be that ordering (the superscripts just indicate membership in* A). *We define the evaluation function (search power of* A) *as*

$$g(A) := \sum_{k=1}^{|A|} s_k^A \prod_{j=k+1}^{|A|} \left(1 - \frac{s_j^A}{w_j^A} \right).$$

1.2   Outline and Results of the Paper

Multi-source Beachcombers' Problems t-SBP, when compared to 1-SBP, add the additional algorithmic complication of partitioning to the existing scheduling problem. In Sect. 2, we explore how the added complexity of partitioning makes the problem NP-hard. In contrast, we prove that some instances of t-SBP admit polynomial time solutions.

Theorem 1. *It is NP-hard to find the optimal solution for arbitrary instances of 2-SBP, while some of the hard instances of 2-SBP admit a* $1/(1 + \epsilon^2)$-*approximation which requires* $O(\frac{n^3 \log m}{\epsilon})$ *steps, where* $m = \max(1/(1 - s_i))$. *The solution for 1-SBP is a* $\frac{1}{2}$-*approximation for 2-SBP, and deterministic 0.5569-approximation for 2-SBP is also possible that requires* $O(n \log n)$ *many computational steps.*

Theorem 2. *Greedy-Partition (Algorithm 3) solves instances of* t-SBP *with* S-*uniform swarms optimally in* $O(n \log n)$ *many steps.*

We then turn our attention to general t-SBP problems. Based on results for 1-SBP established in [1], in Sect. 3 we explore a very efficient and elegantly simple randomized algorithm (Algorithm 4) with constant approximation:

Theorem 3. *There exists a randomized algorithm for* t-SBP *with expected approximation ratio* $1 - \left(1 - \frac{1}{t}\right)^t$ *which needs* $O(n \log(t))$ *random bits, and runs in* $O(n \log n)$ *many steps.*

1.3   Related Work

Traditional graph search originates with the work of Koopman [2], who defined the goal of a searcher as minimizing the time required to find a target object. The work of Stone [3] focuses on the problem of optimizing the allocation of effort (by the searcher) required to search for a target. The author takes a Bayesian approach, assuming there is a prior distribution for the target location (known to the searcher) as well as a function relating the conditional probability of detecting a target given it is located at a point, to the effort applied. In the game theoretic approach studied in [4], the graph *exploration* problem is that of designing an algorithm for the agent that allows it to visit all of the nodes and/or edges of the network. Coupled with this problem is when autonomous, co-operative mobile agents are *searching* for a particular node or item in the network; a problem originating in the work of Shannon [5]. These questions, and similar problems including *swarming, cascading, community formation*, are common in the study of the Internet, P2P, information and social networks.

The Ants Nearby Treasure Search (ANTS) problem [6], though different, is somewhat related to our study. In this problem, k identical mobile robots, all beginning at a start location, are collectively searching for a treasure in the two-dimensional plane. The treasure is placed at a target location by an adversary, and the goal is to find it as fast as possible (as a function of both k and D, where D is the distance between the start location and the target). This is a generalization of the search model proposed in [7], in which the cost of the search is proportional to the distance of the next probe position (relative to the current position) and the goal is to minimize this cost. Related is the w-lane cow-path problem (see [8,9]), in which a cow is standing at a crossroads with w paths leading off into unknown territory. There is a grazing field on one of the paths, while the other paths go on forever. Further, the cow won't know the field has been found until she is standing in it. The goal is to find the field while travelling the least distance possible. Layered graph traversal, as investigated by [10,11], is similar to the cow-path problem, however it allows for short-cuts between paths without going through the origin. Research in [12] is concerned with exploring m concurrent semi-lines (rays) using a single searcher, where a potential target may be located in each semi-line. The goal is to design efficient search strategies for locating t targets (with $t \leq m$). Another model studied in [13] introduces a notion of speed of the agents to study the gathering problem, in which there is an unknown number of anonymous agents that have values they should deliver to a base station (without replications).

The Multi-source Beachcombers' Problem is a combination of two tasks: *scheduling* and *partitioning*. Scheduling jobs with non-identical capacity require-ments or sizes, on single batch processing, to minimize total completion time and makespan, as well as variants of this problem, are studied in several papers including [14–16] and the survey paper [17]. However, they all differ from our investigations in that they do not consider the interplay and trade-offs between walking and searching. It is the partitioning aspect that seems to account for the hardness of t-SBP. This aspect of t-SBP can be reduced to the problem of grouping n items into m subsets $S_1, \ldots, S_m$ to minimize an objective function $g(S_1, \ldots, S_m)$. This is a well-studied problem (c.f. [18–20]), with applications in operations research for inventory control.

From an algorithmic perspective, our work is closely related to Set Partition Problems with Additive Objective (SPAO), see also Sect. 2. A special case of these problems are the so-called Combinatorial Allocation Problems (CAP), where one is given a function $g : 2^{[n]} \mapsto \mathbb{R}_+$ and a fixed integer t, and the objective is to maximize $\sum_{i=1}^{t} g(A_i)$ over all partitions $A_1, \ldots, A_t$ of $[n]$. We will later prove that t-SBP is a SPAO and in fact a CAP where the function g is sub-modular. A $1 - (1 - 1/t)^t$ approximation algorithm is already known [21] for CAPs (matching our performance of Theorem 7). More recently, Feige and Vondrák [22] showed that the same problem is APX-hard, and they improved upon [21] (for large enough values of t) by presenting a $1 - 1/e + \epsilon$ approximation algorithm for $\epsilon \approx 10^{-5}$. Both these are randomized LP-based algorithms, which utilize the solution of an LP with $t \cdot n$ many variables in order to allocate at random the n robots to the t locations (using the language of our problem). As a result, the running time of the previous algorithms is dominated by the running time to solve this LP. In contrast, we propose a much simpler (and oblivious) randomized allocation rule that can be implemented in linear time, and that achieves a $1 - (1 - 1/t)^t$ approximation in the general case. To achieve this, we heavily rely on the special structure of our function g, which also allows us to establish improved approximations, or even exact solutions, for more restricted yet interesting variations of our problem.

2 t-SBP as a Partitioning Problem and its Hardness

We begin our examination of t-SBP by considering the problem where robots are already partitioned into groups $A_1, \ldots, A_t$. It is not difficult to see that (as in 1-SBP) we have the same necessary conditions of optimality, i.e. robots start searching at the same time, they all finish at the same time, etc. In other words, having fixed a partition, the optimal solution can be found by solving t many 1-SBP instances. In particular, each group of robots will process a subinterval that is proportional to its search power. This observation reduces the problem of finding an optimal schedule for t-SBP into the problem of correctly guessing a partition of the swarm into t groups. With this in mind, one may recall the following family of well-studied problems [20].

Definition 2 (The Set Partition Problem with Additive Objective (SPAO)). *Consider n items, a function $g : 2^{[n]} \mapsto \mathbb{R}^+$, and a fixed integer t. In the Set Partition Additive Objective (SPAO) problem we are looking for a partition of $[n]$ into t disjoint sets $A_1, \ldots, A_t$ that maximizes $\sum_{i=1}^{t} g(A_i)$.*

Note that t-SBP problems are almost a generalization of SPAO, since our objects now have 2-dimensional attributes, i.e. the walking and the searching speeds. We can translate the Beachcombers' Problem into a partitioning problem resembling SPAO:

Observation 4 (t-SBP Revisited). *Consider n items, each having two attributes s_i, w_i, with $s_i < w_i$. Define the evaluation function $g : 2^{[n]} \mapsto \mathbb{R}^+$ as the search power of subsets of the swarm with the same attributes (See Definition 1). Then, in the t-SBP problem we are looking for a partition of $[n]$ into t disjoint sets $A_1, \ldots, A_t$ that maximizes $\sum_{i=1}^{t} g(A_i)$.*

One may consider a greedy approach where we first order the items by non-decreasing w_i's, and then assign robot r into set A_j, chosen from $A_1, \ldots, A_t$, where $g(A_j)$ is minimum (before adding r). This idea is equivalent to scheduling a robot to the group which minimizes the amount it walks. Note that, as described by the solution to 1-SBP given in Lemma 2, adding item (robot) r to A_j, increases the value of the objective by

$$c_r = s_r \left(1 - \frac{1}{w_r} g(A_j) \right). \tag{2}$$

We now recall from the literature a combinatorial problem that plays a crucial role in our upper and lower bounds. In PRODUCTPARTITION we are given positive integers $m_1, \ldots, m_n$ (with $m_i > 1$) and asked if there is a partition of them, A, B, such that $\prod_{i \in A} m_i = \prod_{i \in B} m_i$. PRODUCTPARTITION has a natural optimization version, MINMAXPRODUCTPARTITION, where we try to minimize $\max\{\prod_{i \in A} m_i, \prod_{i \in B} m_i\}$ over all partitions A, B of the n integers. As Ng et al. showed [23], this objective function can be well approximated, while the decision problem is hard. We use this to establish the hardness of 2-SBP for W-uniform swarms by reducing from PRODUCTPARTITION.

Theorem 5. *It is NP-hard to solve 2-SBP with a W-uniform swarm.*

Proof. From the hardness result of PRODUCTPARTITION, it suffices to reduce from PRODUCTPARTITION. Our reduction considers a W-uniform swarm and sets the searching speeds as $s_i = 1 - \frac{1}{m_i}$. We argue that if we can solve 2-SBP in polynomial time, then we can decide PRODUCTPARTITION in polynomial time as well.

Given a partition A, B of the W-uniform swarm, we know the best schedule has cost

$$2 - \prod_{i \in A}(1 - s_i) - \prod_{i \in B}(1 - s_i),$$

given by Definition 1 (Search Power) and the solution given in [1]. Hence, optimizing the finishing time for this instance of 2-SBP is equivalent to finding a partition A, B so as to minimize

$$\prod_{i \in A}(1 - s_i) + \prod_{i \in B}(1 - s_i).$$

Since $\prod_{i \in A \cup B}(1 - s_i)$ is invariant for all partitions, the problem can be translated into minimizing

$$\left| \prod_{i \in A}(1 - s_i) - \prod_{i \in B}(1 - s_i) \right| = \left| \prod_{i \in A} \frac{1}{m_i} - \prod_{i \in B} \frac{1}{m_i} \right|.$$

Thus, optimizing the schedule of the instance finds a partition that solves PRO-
DUCTPARTITION, if one exists.

Theorem 6. *W-Uniform* 2-SBP *instances* $s_1, \ldots, s_n$ *for which* $\frac{1}{1-s_i} \in \mathbb{N}$, *admit
a* $1/(1 + \epsilon^2)$-*approximation which requires* $O(\frac{n^3 \log m}{\epsilon})$ *steps, where* $m = \max(1/(1 - s_i))$.

Proof. The reader might expect that we will use the FPTAS for MINMAXPRO-
DUCTPARTITION as a subroutine. Actually, we do so in the crudest possible way!
Given a swarm with respective searching speeds $s_1, \ldots, s_n$ with $s_i = 1 - \frac{1}{m_i}$, $m_i \in \mathbb{N}$, we run the FPTAS for MINMAXPRODUCTPARTITION for input $m_1, \ldots, m_n$,
to obtain a partition $A \subseteq \{1, \ldots, n\}$. Let this be an $\alpha := 1 + \epsilon$ approximation to
the MINMAXPRODUCTPARTITION problem.

   We use this partition to split the robots between the two endpoints, and we
schedule each group optimally as described in the solution from [1]. For the ease
of notation, we set $m^A := \prod_{i \in A} m_i$ and $M := \prod_{i=1}^n m_i$.

   Instead of maximizing the combined searching speed of the groups, we fix
the segment to be explored as unit length and (equivalently) wish to minimize
the finishing time of the search schedules. Therefore, the cost T of our proposed
algorithm and the cost of the optimal solution T_{OPT} are

$$T = \left(2 - \frac{1}{m^A} - \frac{m^A}{M}\right)^{-1} \quad \text{and} \quad T_{\text{OPT}} = \left(2 - \frac{1}{m^D} - \frac{m^D}{M}\right)^{-1}, \qquad (3)$$

where D is the optimal underlying partition of the best scheduling. We claim
that $T/T_{\text{OPT}} \leq 1 + \epsilon^2$.

   In order to break the symmetry, we assume that

$$m^A \geq \sqrt{M} \quad \text{and} \quad m^D \geq \sqrt{M}. \qquad (4)$$

   The key observation towards concluding our claim is that the partition D
which minimizes

$$\left(2 - \frac{1}{m^D} - \frac{m^D}{M}\right)^{-1}$$

is the same as the one which minimizes

$$\max\{m^D, \frac{M}{m^D}\} \overset{(4)}{=} m^D.$$

Since we run the FPTAS for MINMAXPRODUCTPARTITION, the cost of the
solution was $\max\{m^A, \frac{M}{m^A}\} \overset{(4)}{=} m^A$, and in particular

$$m^A \leq \alpha \, m^D. \qquad (5)$$

We are now ready to compute the approximation guarantee of our proposed
algorithm:

$$\frac{T}{T_{\text{OPT}}} \overset{(3)}{=} \frac{2 - \frac{1}{m^D} - \frac{m^D}{M}}{2 - \frac{1}{m^A} - \frac{m^A}{M}} \overset{(5),(4)}{\leq} \frac{2 - \frac{1}{m^D} - \frac{m^D}{M}}{2 - \frac{1}{\alpha\, m^D} - \frac{\alpha\, m^D}{M}}$$

$$\overset{(4)}{\leq} \frac{2 - \frac{2}{m^D}}{2 - \frac{1}{\alpha\, m^D} - \frac{\alpha}{m^D}} = \frac{2m^D - 2}{2m^D - \left(\frac{1}{\alpha} + \alpha\right)} \leq \frac{2}{4 - \left(\frac{1}{\alpha} + \alpha\right)},$$

where the last inequality is due to the fact that the ratio decreases with D. Plugging in $\alpha = 1 + \epsilon$ it is easy to see then that last ratio is no more than $1 + \epsilon^2/2$.

2.1 A Deterministic Approximation Guarantee for 2-SBP

In this section we show that the solution to 1-SBP is a trivial $\frac{1}{2}$-approximation for 2-SBP, and present a deterministic 0.5569-approximation for 2-SBP. The algorithm is based on the optimal solution to 1-SBP and a shifting mechanism that trades the walking time of robots for longer searching times. As before, we assume that the interval to be searched is $[0,1]$. If this is not the case, the solution is easily scaled.

Consider the optimal solution to 1-SBP in which robots $1, ..., n$ are ordered according to their respective search intervals $c_1, ..., c_n$, where $\sum_{i=1}^n c_i = 1$. We denote by $d_1, ..., d_n$ distances in the optimal solution that robots walk towards their respective search intervals. In particular, $d_1 = 0$, and $d_i = \sum_{j=1}^{i-1} |c_j|$, for all $1 < i \leq n$.

Lemma 3. *The robots $1, ..., i$ with search intervals $c_1, ..., c_i$ and which search a combined interval of length d_{i+1} in the optimal solution to 1-SBP cannot search an interval of length greater than $d_i + d_{i+1}$ in 2-SBP.*

Proof. Follows from the fact that even if robots $i - 1$ and i operate at different ends of the searched segment in 2-SBP the total contribution at the respective ends cannot exceed values d_i and d_{i+1}. Otherwise these robots would have to be preceded by some extra robots and that would violate the property of the optimal solution in which participating robots must be ordered according to their walking speeds.

Corollary 1. *The solution for 1-SBP provides a $\frac{1}{2}$-approximation for 2-SBP.*

Proof. Follows immediately from Lemma 3 when considering all robots.

Consider any robot i. In the optimal solution to 1-SBP, robot i walks a fixed distance d_i from the front (left end) of the searched segment to its search interval of length x_i. We observe that the length of the searched interval x_i could be longer if i walked a shorter distance. In particular, if i walked distance $\beta \cdot d_i$ instead, for some constant $0 < \beta < 1$, the searched interval can be extended by $y_i = \frac{d_i}{w_i}(1 - \beta)s_i$, where $\frac{d_i}{w_i}$ refers to the walking time of i in the optimal solution to 1-SBP. Since w_i, s_i and d_i are fixed, the extension y_i is a linear function of β. We say that a group X of searched intervals is subject to β-expansion if for each $x_i \in X$, the respective robot i walks a distance at most βd_i from the front of the searched segment. The total gain (extra search distance) of β-expansion is defined as $\sum_{x_i \in X} y_i$. This is also a linear function of β.

We now present our 0.5569-approximation algorithm to 2-SBP, which we will call Swap-and-Expand. For convenience, we have the two groups of robots walking in opposing directions starting from the two endpoints of the interval. The proof of correctness of our algorithm completes the proof of Theorem 1.

Algorithm 2. Swap-and-Expand

Require: swarm $(\mathbf{s}, \mathbf{w})$
Ensure: Schedule of the swarm.
1: Sort the robots in non-decreasing walking speed.
2: Calculate the search intervals $(c_1, \ldots, c_n)$ for Algorithm Comb.
3: Find i such that $\sum_{k=1}^{i-1} c_k < 1/2 \leq \sum_{k=1}^{i} c_k$
4: Assign robots $1 \ldots i$ to group A_0 and robots $i+1 \ldots n$ to group A_1
5: Place group A_0 at point 0 with direction *right* and group A_1 at point 1 with direction *left*.
6: Execute Algorithm Comb with the robots in A_0, and concurrently with the robots in A_1.

The algorithm is simple: either moving the robots leads to a significant improvement (large β-expansion), or it does not. In the latter case, we can show that the optimal solution for 1-SBP is better than a $\frac{1}{2}$-approximation. We observe that the output from this algorithm solves 2-SBP at least as well as simply using the solution for 1-SBP.

Lemma 4. *Swap-and-Expand gives an α-approximation for* 2-SBP, *where $\alpha = \frac{1}{8}\left(13 - \sqrt{73}\right) > 0.5569$.*

Proof. For some $0 \leq \delta < 1/2$, consider two cases referring to the total gain of a $(1/2 + \delta)$-expansion of A_1: **Case 1** where the total gain is $\leq \delta$ and the complementary **Case 2**.

Case 1. In this case the total gain of $(1/2 + \delta)$-expansion is at most δ. We first observe that in this case (the best possible) 0-expansion would give the total gain $\leq \frac{\delta}{1/2-\delta}$ (follows from linearity of the expansion mechanism). In other words, even if all intervals in A_1 were moved to the beginning of the searched segment (i.e., respective robots did not have to walk at all) the total gain would not exceed $\frac{\delta}{1/2-\delta}$. Concluding, one can observe that the total length of the search intervals in any 2-SBP solution would not exceed $d_i + d_{i+1} + \sum_{j=i+1}^{n} |c_j| + \frac{\delta}{1/2-\delta}$. This is equal to $d_i + 1 + \frac{\delta}{1/2-\delta}$, and since $d_i < 1/2$ we get the bound $\frac{1}{\frac{3}{2} + \frac{\delta}{1/2-\delta}} = \frac{2-4\delta}{3-2\delta}$.

Case 2. In this case the total gain of the $(1/2+\delta)$-expansion is larger than δ. This gain will be used by the approximation algorithm in the 2-SBP setting where one is allowed to allocate robots with their search intervals at the two ends of the searched segment. We propose the following solution. From the optimal solution to 1-SBP we remove intervals $c_1, \ldots, c_i$. These will be eventually placed in the reversed order at the right end of the searched segment in the 2-SBP setting. This removal leaves a gap of size d_{i+1} at the left end of the searched segment. We shift all remaining intervals (without changing their order) in A_1 maximally towards the left end of the searched segment avoiding overlaps.

(a) If all segments in A_1 experience $(1/2 + \delta)$-expansion, i.e., if each $c_i \in A_1$ shortens its distance d_i to the left end of the searched segment to $\leq d_i \cdot (1/2 + \delta)$ the total gain is $\geq \delta$. In this case the interval searched in time t (the optimal time for 1-SBP) in 2-SBP is of length at least $1 + \delta$. Since the optimal solution in 1-SBP gives $\frac{1}{2}$-approximation in 2-SBP, we have the approximation ratio $\frac{1+\delta}{2}$.

(b) Otherwise, if at least one segment (take the one with the smallest index k) $c_k \in A_1$ does not experience $(1/2 + \delta)$-expansion, its left end is at distance $> d_k \cdot (1/2 + \delta)$ from the left end of the searched segment. In this case the total gain obtained by $c_{i+1}, \ldots, c_{k-1}$ is $\geq d_k \cdot (1/2 + \delta) - \sum_{j=i+1}^{k-1} |c_j|$. Now since $\sum_{j=i+1}^{k-1} |c_j| \leq d_k - 1/2$ (since c_i contains the point $1/2$) we conclude that the total gain is $\geq d_k \cdot (1/2 + \delta) - d_k + 1/2 = d_k \cdot (\delta - 1/2) + 1/2$. Since $d_k \leq 1$ the total gain is at least δ. As in **Case 2a**, we get the approximation ratio $\frac{1+\delta}{2}$.

To this point, we have said nothing about the value of δ. However, given an arbitrary swarm either case may hold. Therefore, we want to choose a value of δ which maximizes $\min\left(\frac{2-4\delta}{3-2\delta}, \frac{1+\delta}{2}\right)$. Since the first function is monotonic decreasing (with respect to δ) and the second is monotonic increasing, $\min\left(\frac{2-4\delta}{3-2\delta}, \frac{1+\delta}{2}\right)$ is maximized when $\frac{2-4\delta}{3-2\delta} = \frac{1+\delta}{2}$. We solve this for δ and find its roots, only one of which falls in the interval $[0, 1]$, when $\delta = \frac{9-\sqrt{73}}{4}$, from which we obtain an approximation ratio bounded below by $\frac{1}{8}\left(13 - \sqrt{73}\right) \geq 0.5569$.

2.2 An Exact Solution for t-SBP with S-Uniform Swarms

We now present Greedy-Partition (Algorithm 3) and show that it solves instances of t-SBP with S-uniform swarms in $O(n \log n)$ many steps (as stated in Theorem 2). We assume that the interval to be searched is $[0, 1]$. Again, if this is not the case, the solution is easily scaled.

Algorithm 3 works by sorting the robots, separating them into groups in a revolving manner, calculating the search power of the groups, and placing the groups at positions on the interval so that by executing the algorithm for 1-SBP, each group finishes searching their respective subintervals at the same time.

Proof (Theorem 2). It is clear that the algorithm runs in $O(n \log n)$ time - the time required to sort the robots. We note that the robots are passed in sorted order to Comb, so that it runs in $O(1)$ time.

It now remains to show that Algorithm 3 is optimal. We begin by noting that no two of the t slowest walking robots will be placed in the same group, and therefore none of them will walk at all. Since they all search at the same rate, they are interchangeable and any solution which has any permutation of the first t robots, but identical otherwise will be equivalent to the solution provided by Algorithm 3.

Algorithm 3. Greedy-Partition

Require: swarm $(\mathbf{s}, \mathbf{w})$, integer t.
Ensure: Schedule of the swarm for t-SBP.
1: Sort the robots in non-decreasing walking speeds.
2: **for** $i = 1, \ldots, n$ **do**
3:    Assign robot i to group $A_{(i \bmod t)}$
4: **end for**
5: **for** $i = 1, \ldots, t$ **do**
6:    $x_j := \sum_{k=1}^{|A_i|} \prod_{j=k+1}^{|A_i|} \left(1 - \frac{1}{w_j^{A_i}}\right)$
7: **end for**
8: **for** $i = 1, \ldots, t$ **do**
9:    Place group A_i at point $\frac{\sum_{k=1}^{i-1} x_k}{\sum_{k=1}^{t} x_k}$ with search direction *right*, A_1 is placed at point
      0.
10: **end for**
11: **for** $i = 1, \ldots, t$ (in parallel) **do**
12:    Execute Algorithm `Comb` with the robots in A_i.
13: **end for**

Assume then, that there exists an optimal solution which is not equivalent to the solution provided by Algorithm 3. In this case, we can find two robots $a \in A_j$, $b \in A_k$ such that $\omega_a < \omega_b$ and $w_a > w_b$ where ω_i the interval walked by robot i. In other words, there exist two robots where the faster walker walks a shorter distance that the slower one. We first observe that:

Lemma 5. *Swapping a and b increases the length searched by a and b combined.*

Proof. Since both robots have identical searching speeds, we can simply consider the total time walking - as the length of the combined interval searched will be maximized when the combined time walking is minimized. Both before and after the swap, each robot must walk at least ω_a. Before the swap, a begins searching immediately while b must walk a further $\omega_b - \omega_a$ before it can start searching. So the total time spent walking by both robots (again ignoring the first interval) is $\frac{\omega_a}{w_a} + \frac{\omega_a}{w_b} + \frac{\omega_b - \omega_a}{w_b}$. Similarly, the total time spent walking by both robots after the swap is $\frac{\omega_a}{w_a} + \frac{\omega_a}{w_b} + \frac{\omega_b - \omega_a}{w_a}$. Since $w_a > w_b$, the total time walking is less after the swap (Fig. 1).

Therefore, swapping a and b must negatively affect other robots, otherwise we have reached a contradiction. Since it does not affect any robots that search intervals closer to their source points than a and b, then there exist robots which search intervals farther from their source points than the intervals searched by a and b. We can assume that the robots searching these outer intervals are arranged optimally, as described by Algorithm 3. If they are not, then there exist two robots farther from their source points than a and b which have the same properties as a and b, and so we can consider them instead.

We now consider how swapping a and b affects the outer intervals searched by the remaining robots. We denote the intervals searched by robot i before and

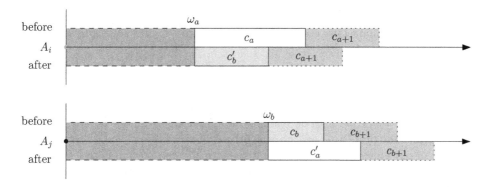

Fig. 1. Effect of swapping a and b on outer intervals

after the swap as c_i and c'_i, respectively. We assume that $w_a + x_a < w_b + x_b$. If this is not true, then we also swap all robots which search intervals after a and b, and the rest of the following argument will hold. After the swap, the robots which originally started after a will walk less than they did before the swap. Similarly, the robots after b will walk more than they did before the swap. We note that $a + 1$ walked a shorter distance than $b + 1$ before the swap and so we say that this swap *unbalances* the outer intervals, as the ones that were closer to their start point before the swap become even closer at the expense of the others, which move further away from their start point by a corresponding amount.

Lemma 6. *Unbalancing the outer intervals improves the search time.*

Proof. Suppose that the intervals are unbalanced by $\epsilon > 0$. Consider the robots $a + 1$ and $b + 1$ which originally follow both a and b. We observe that as a pair, after the swap, the time it takes them to search intervals of the same combined distance as their original intervals improves by $\frac{\epsilon}{w_{a+1}} - \frac{\epsilon}{w_{b+1}}$. Since $w_{a+1} < w_{b+1}$ we know this is a positive amount. Similarly, we can show that the pairs $(a + 2, b + 2), \ldots, (a + k, b + k)$ all improve their search time by the imbalance. We know that there must be at least as many robots which follow a as there are that follow b - since the distribution of these robots follows Comb.

A careful examination of the swap of a and b will reveal that it does not create a strict imbalance as described above - the one set of intervals shift closer to their start point less than the other set shifts farther from theirs. This shift also has the effect of inflating the intervals shifted closer, so that the first few intervals will be shifted more than the last few. These effects may at first seem to mean that the above argument cannot be applied, however we note that the difference in the amount that the intervals shift is exactly the amount that was gained by swapping a and b. Therefore, if we first consider the swap without the increased search efficiency, we can see that the imbalance improves the search time. If we then add in the improvement, it will again improve the search time.

We have therefore shown that no matter where a and b are positioned, we can improve the solution. This contradicts our assumption, and so Algorithm 3 is optimal.

3   A Randomized Algorithm for t-SBP

In this section, we propose an oblivious randomized algorithm for t-SBP. We fix some input to the problem, i.e. a swarm with attributes $(s_1, w_1), \ldots, (s_n, w_n)$. For the sake of our analysis, we assume that the swarm is ordered in non-decreasing walking speeds w_i. Denote the optimal solution to t-SBP for the above instance by $\mathrm{OPT}_t(\mathbf{s}, \mathbf{w})$. We propose the next randomized algorithm. We note that Algorithm 4 is identical to Algorithm 3 except for how the robots are partitioned into groups.

Algorithm 4. Oblivious Randomized (OR) Algorithm

Require: swarm $(\mathbf{s}, \mathbf{w})$, integer t and $O(n \log(t))$ random bits.
Ensure: Schedule of the swarm for t-SBP.
1: **for** $i = 1, \ldots, n$ **do**
2:     Choose k independently and uniformly at random from $1, \ldots, t$.
3:     Assign robot i to group A_k
4: **end for**
5: **for** $i = 1, \ldots, t$ **do**
6:     $x_j := \sum_{k=1}^{|A_i|} \prod_{j=k+1}^{|A_i|} \left(1 - \frac{1}{w_j^{A_i}} \right)$
7: **end for**
8: **for** $i = 1, \ldots, t$ **do**
9:     Place group A_i at point $\frac{\sum_{k=1}^{i-1} x_k}{\sum_{k=1}^{t} x_k}$ with search direction *right*, A_1 is placed at point 0.
10: **end for**
11: **for** $i = 1, \ldots, t$ (in parallel) **do**
12:     Execute Algorithm `Comb` with the robots in A_i.
13: **end for**

Lemma 7. *The expected value of the Oblivious Randomized algorithm for t-SBP on input $(\mathbf{s}, \mathbf{w})$ equals $t \cdot \mathrm{OPT}_1 (s/t, \mathbf{w})$.*

Proof. The Oblivious Randomized algorithm assigns each robot to one of the t groups uniformly at random. Denote the induced partition by $A_1, A_2, \ldots, A_t$. By symmetry, and due to linearity of expectation, the expected value of the Oblivious Randomized algorithm is t times the expected searched interval of the group A_1.

Let now B_r denote the random variable that equals the search power of group A_1, when we only use the first r robots with respect to their walking speeds. Clearly, $\mathbb{E}[B_n]$ is the expected search power of A_1, and due to the short discussion above, the expected value of our randomized algorithm is $t \cdot \mathbb{E}[B_n]$.

In order to compute $\mathbb{E}[B_n]$ we set up a recursion. We observe that

$$\mathbb{E}[B_n] = \mathbb{P}[n \in A_1]\,\mathbb{E}[B_n \mid n \in A_1] + \mathbb{P}[n \notin A_1]\,\mathbb{E}[B_n \mid n \notin A_1] \tag{6}$$

$$= \frac{1}{t}\,\mathbb{E}[B_n \mid n \in A_1] + \frac{t-1}{t}\,\mathbb{E}[B_n \mid n \notin A_1] \tag{7}$$

$$= \frac{1}{t}\,\mathbb{E}\left[B_{n-1} + s_n\left(1 - \frac{1}{w_n}B_{n-1}\right)\right] + \frac{t-1}{t}\,\mathbb{E}[B_{n-1}] \tag{8}$$

$$= \frac{1}{t}s_n + \left(1 - \frac{s_n}{tw_n}\right)\mathbb{E}[B_{n-1}], \tag{9}$$

where line (6) is due to the fact that the fastest walking robot is either in or outside A_1, line (7) is from the fact that robots are assigned to groups uniformly at random, line (8) is due to (2) and finally (9) is because of linearity of expectation.

We conclude the lemma by recalling that if X_n denotes the optimal solution for one group when we use the first n robots, then $X_n = s_n + \left(1 - \frac{s_n}{w_n}\right)X_{n-1}$ (as it can be easily derived from the closed form solution given in [1]). Hence, $\mathbb{E}[B_n]$ satisfies the exact same recurrence, but searching speeds are scaled by t.

By Lemma 7, the expected performance of the OR Algorithm is exactly

$$\frac{t \cdot \mathrm{OPT}_1\left(\frac{\mathbf{s}}{t}, \mathbf{w}\right)}{\mathrm{OPT}_t(\mathbf{s}, \mathbf{w})}, \tag{10}$$

from which we obtain naive bounds of $\mathrm{OPT}_1\left(\frac{\mathbf{s}}{t}, \mathbf{w}\right) \geq \frac{1}{t}\mathrm{OPT}_1(\mathbf{s}, \mathbf{w})$, and that $\mathrm{OPT}_t(\mathbf{s}, \mathbf{w}) \leq t \cdot \mathrm{OPT}_1(\mathbf{s}, \mathbf{w})$.

The approximation ratio is then $1/t$, which is the ratio we achieve if we assign all robots to one group. The expected performance should be better, given that we assign robots to random groups. The crux of our analysis is based on the trade-offs between the bounds we have for $\mathrm{OPT}_1\left(\frac{\mathbf{s}}{t}, \mathbf{w}\right)$ and $\mathrm{OPT}_t(\mathbf{s}, \mathbf{w})$, with respect to $\mathrm{OPT}_1(\mathbf{s}, \mathbf{w})$. We now show a restatement of Theorem 3.

Theorem 7. *For the t-SBP problem on instance $(\mathbf{s}, \mathbf{w})$, the OR Algorithm outputs (in expectation) a schedule of combined search power at least*

$$(1 - (1 - 1/t)^t)\mathrm{OPT}_t(\mathbf{s}, \mathbf{w}).$$

The trick in our analysis is to first find the worst configuration of swarms that minimizes (10) over all swarms that satisfy $\mathrm{OPT}_t(\mathbf{s}, \mathbf{w}) = \alpha \cdot \mathrm{OPT}_1(\mathbf{s}, \mathbf{w})$ for some $1 \leq \alpha \leq n$. Effectively, this determines the worst approximation ratio, call it $R(\alpha)$, for the Oblivious Randomized Algorithm, with the restriction that $\mathrm{OPT}_t(\mathbf{s}, \mathbf{w}) = \alpha \cdot \mathrm{OPT}_1(\mathbf{s}, \mathbf{w})$. Finding $\max_{1 \leq \alpha \leq n} R(\alpha)$ will then determine the desired bound on the approximation ratio (as described in Theorem 7). In what follows, we focus on estimating $R(\alpha)$.

Lemma 8. *The performance of the Oblivious Randomized Algorithm, $R(\alpha)$, is worst when for each robot i, we have $s_i \approx w_i$.*

Proof. Since $t \geq 2$, there exists a group in $\mathrm{OPT}_t(\mathbf{s}, \mathbf{w})$ in which a robot r does not walk (only searches), but walks in $\mathrm{OPT}_1(\mathbf{s}, \mathbf{w})$. Note that the value w_r appears only the numerator of (10). Since in particular we have

$$\mathrm{OPT}_1 \left(\frac{\mathbf{s}}{t}, \mathbf{w} \right) = \sum_{k=1}^{n} \frac{s_k}{t} \prod_{j=k+1}^{n} \left(1 - \frac{s_j}{tw_j} \right),$$

it is clear that the ratio is minimized when w_r is minimized. On the other hand, $w_r > s_r$, and hence when the approximation ratio is minimized, w_r is infinitesimally close to s_r.

Given that $s_r \approx w_r$, we observe then that $\mathrm{OPT}_1(\mathbf{s}, \mathbf{w}) \approx s_n$, where n is the index of the robot with the fastest walking speed. Effectively, this shows that the approximation ratio we want to lower bound is

$$\frac{t \cdot \mathrm{OPT}_1 \left(\frac{\mathbf{s}}{t}, \mathbf{w} \right)}{\mathrm{OPT}_t (\mathbf{s}, \mathbf{w})} \approx \frac{t \cdot \mathrm{OPT}_1 \left(\frac{\mathbf{s}}{t}, \mathbf{w} \right)}{\alpha \cdot s_n},$$

with the restriction that $\mathrm{OPT}_t(\mathbf{s}, \mathbf{w}) = \alpha \cdot s_n$. But, we further observe that the rest of the walking speeds $w_2, \ldots, w_n$ appear only in the numerator of the ratio we want to bound, and (identically to our argument for robot r), the ratio is smallest when the w_i attain as small values as possible. This shows that $s_i \approx w_i$ for $i = 2, \ldots, n$, while the walking speed w_1 of robot 1 is irrelevant to $\mathrm{OPT}_1(\mathbf{s}, \mathbf{w})$.

We can now express $R(\alpha)$ as the minimum of a relatively easy rational function, after which we will have all we need to prove Theorem 3.

Lemma 9. *The instance that minimizes the performance $R(\alpha)$ of the Oblivious Randomized Algorithm for t-SBP is a swarm with searching speeds $s_1 \leq s_2 \leq \ldots \leq s_t$. Moreover, the value $R(\alpha)$ is given as the infimum of the following optimization problem*

$$\min_{s_1, \ldots, s_t} \frac{\sum_{k=1}^{t} s_k \left(1 - \frac{1}{t} \right)^{t-k}}{\alpha \cdot s_t} \quad subject\ to \quad \sum_{k=1}^{t} s_k = \alpha \cdot s_t$$

Proof. For a swarm $(\mathbf{s}, \mathbf{w})$ that minimizes $R(\alpha)$, let $i_1, \ldots, i_t$ be the indices of robots that search the last intervals (i.e. start searching the latest in time) for the t different groups of $\mathrm{OPT}_t (\mathbf{s}, \mathbf{w})$. Below we argue that the exact same robots are scheduled last when computing $\mathrm{OPT}_1 \left(\frac{\mathbf{s}}{t}, \mathbf{w} \right)$.

Since robots in each group are always ordered in non-decreasing walking speeds, it also follows that each of the robots $i_1, \ldots, i_t$ is the fastest in each of the t groups. By Claim 8, we know that we may assume that all walking speeds almost coincide with the searching speeds, and hence

$$\mathrm{OPT}_t (\mathbf{s}, \mathbf{w}) \approx \sum_{s=1}^{t} s_{i_s}. \tag{11}$$

However, by Claim 8 for the configuration that determines $R(\alpha)$ we also have

$$t \cdot \text{OPT}_1\left(\frac{\mathbf{s}}{t}, \mathbf{w}\right) \approx \sum_{k=1}^{n} s_k \left(1 - \frac{1}{t}\right)^{n-k}. \tag{12}$$

This immediately implies that $R(\alpha)$ is given as the ratio of expression (12) over expression (11), with the restriction that $\sum_{s=1}^{t} s_{i_s} = \alpha \cdot s_n$. Recall that this corresponds to the ratio for a swarm where the searching speed for every robot is infinitesimally close to its walking speed. Also, (11) not only gives the value for t-SBP, but also tells us that the robots i_s, $s = 1, \ldots, t$ must be the fastest robots in the swarm with respect to their walking speeds (that coincide with the searching speeds). Otherwise, we could improve the value $\text{OPT}_t\left(\mathbf{s}, \mathbf{w}\right)$. But then, the schedule we have for $\text{OPT}_1\left(\frac{\mathbf{s}}{t}, \mathbf{w}\right)$ should also schedule the exact same t robots to search the last t intervals, meaning that the last t robots in the sum (12) must be $i_1, i_2, \ldots, i_t$ (and without loss of generality, as this does not affect the optimal value of t-SBP, we may assume that $s_{i_1} \leq s_{i_2}, \ldots \leq s_{i_t}$). But then, just ignoring the contribution of the first $n - t$ robots, we have that

$$\sum_{k=1}^{n} s_k \left(1 - \frac{1}{t}\right)^{n-k} \geq \sum_{k=n-t+1}^{n} s_k \left(1 - \frac{1}{t}\right)^{n-k} = \sum_{s=1}^{t} s_{i_s} \left(1 - \frac{1}{t}\right)^{t-s} \tag{13}$$

Combining now (13) with (12) shows the lemma.

Proof (Theorem 3). Lemma 9 gives the expected performance $R(\alpha)$ of the Oblivious Randomized Algorithm, when $\text{OPT}_t\left(\mathbf{s}, \mathbf{w}\right) = \alpha \cdot \text{OPT}_1\left(\mathbf{s}, \mathbf{w}\right)$ which we rewrite as

$$\sum_{k=1}^{t-1} \frac{s_k}{s_t} = \alpha - 1 \tag{14}$$

while also, the expected performance is the infimum of the expression

$$\frac{\sum_{k=1}^{t} s_k \left(1 - \frac{1}{t}\right)^{t-k}}{\alpha \cdot s_t} = \frac{1}{\alpha} \cdot \sum_{k=1}^{t-1} \frac{s_k}{s_t} \left(1 - \frac{1}{t}\right)^{t-k} + \frac{1}{\alpha} \tag{15}$$

We note that the value of the above expression is invariant under scaling, so for convenience we may set $s_t = 1$. Next, recall that according to Claim 9, the searching speeds satisfy $s_1 \leq s_2 \leq \ldots \leq s_{t-1} \leq 1$. Since by (14) the speeds s_i sum up to $\alpha - 1$, and observing the monotonicity of the coefficients of s_i, one can easily show that the sum (15) is minimized when $s_1 = s_2 = \ldots = s_{t-1} = \frac{\alpha-1}{t-1}$, which is compatible with our restrictions, since $\alpha \leq t$. Then, we have

$$R(\alpha) \stackrel{(15)}{=} \frac{1}{\alpha}\left(\frac{\alpha-1}{t-1} \sum_{k=1}^{t-1} \left(1 - \frac{1}{t}\right)^{t-k} + 1\right)$$

$$= \frac{1}{\alpha}\left(\frac{\alpha-1}{t-1}\left(t - 1 - t\left(1 - \frac{1}{t}\right)^{t}\right) + 1\right) \tag{16}$$

A final observation is that expression (16) is decreasing in α (for every fixed t). Since α is no more than t (recall that α measures how much more OPT_t is compared to OPT_1), we conclude that the expected approximation ratio of the Oblivious Randomized algorithm is $\min_\alpha R(\alpha) = R(t)$. Substituting $\alpha = t$ in (16) gives us what the Theorem claims.

4   Conclusion and Open Problems

In this paper, we proposed and analysed several algorithms for addressing the Multi-source Beachcombers' Problem. There are several other variants of the problem worth studying. Different domain topologies could be considered. To account for the case of faults, we may want to investigate multiple coverage of the domain. Finally, for the Multi-source Beachcombers' Problem, we observe that some (and perhaps all) instances with a W-uniform swarm are NP-hard. However, all instances with S-uniform swarms are solvable in polynomial time. There seems to be some underlying relationship between searching and walking speeds which determines the difficulty of the problem. It would be interesting to explore this further, to see if this relationship can be quantified, and ideally identify the threshold for which the problem tips from easy to hard.

References

1. Czyzowicz, J., Gasieniec, L., Georgiou, K., Kranakis, E., MacQuarrie, F.: The Beachcombers' problem: walking and searching with mobile robots. In: Halldórsson, M.M. (ed.) SIROCCO 2014. LNCS, vol. 8576, pp. 23–36. Springer, Heidelberg (2014)
2. Koopman, B.O.: Search and screening. Operations Evaluation Group,Office of the Chief of Naval Operations, Navy Department (1946)
3. Stone, L.D.: Theory of Optimal Search. Academic Press, New York (1975)
4. Alpern, S., Gal, S.: The Theory of Search Games and Rendezvous, vol. 55. Kluwer Academic Publishers, Dordrecht (2002)
5. Shannon, C.E.: Presentation of a maze-solving machine. In: 8th Conference of the Josiah Macy Jr. Found (Cybernetics), pp. 173–180 (1951)
6. Feinerman, O., Korman, A., Lotker, Z., Sereni, J.S.: Collaborative search on the plane without communication. In: PODC, pp. 77–86. ACM (2012)
7. Baeza-Yates, R.A., Culberson, J.C., Rawlins, G.J.E.: Searching in the plane. Inf. Comput. **106**, 234–234 (1993)
8. Kao, M.Y., Littman, M.L.: Algorithms for informed cows. In: AAAI-97 Workshop on On-line Search (1997)
9. Kao, M.Y., Reif, J.H., Tate, S.R.: Searching in an unknown environment: an optimal randomized algorithm for the cow-path problem. In: SODA, pp. 441–447. Society for Industrial and Applied Mathematics (1993)
10. Fiat, A., Foster, D.P., Karloff, H., Rabani, Y., Ravid, Y., Viswanathan, S.: Competitive algorithms for layered graph traversal. In: FOCS, pp. 288–297. IEEE (1991)
11. Papadimitriou, C.H., Yannakakis, M.: Shortest paths without a map. Theor. Comput. Sci. **84**(1), 127–150 (1991)

12. Angelopoulos, S., López-Ortiz, A., Panagiotou, K.: Multi-target ray searching problems. In: Dehne, F., Iacono, J., Sack, J.-R. (eds.) WADS 2011. LNCS, vol. 6844, pp. 37–48. Springer, Heidelberg (2011)
13. Beauquier, J., Burman, J., Clement, J., Kutten, S.: On utilizing speed in networks of mobile agents. In: PODC, pp. 305–314. ACM (2010)
14. Brucker, P., Gladky, A., Hoogeveen, J.A., Kovalyov, M., Potts, C., Tautenhahn, T., Velde, S.: Scheduling a batching machine. J. Sched. **1**(1), 31–54 (1998)
15. Potts, C., Kovalyov, M.: Scheduling with batching: a review. Eur. J. Oper. Res. **120**(2), 228–249 (2000)
16. Uzsoy, R.: Scheduling a single batch processing machine with non-identical job sizes. Int. J. Prod. Res. **32**(7), 1615–1635 (1994)
17. Allahverdi, A., Ng, C.T., Cheng, T.C.E., Kovalyov, M.: A survey of scheduling problems with setup times or costs. Eur. J. Oper. Res. **187**(3), 985–1032 (2008)
18. Anily, S., Federgruen, A.: Structured partitioning problems. Oper. Res. **39**(1), 130–149 (1991)
19. Chakravarty, A.K., Orlin, J.B., Rothblum, U.G.: Consecutive optimizers for a partitioning problem with applications to optimal inventory groupings for joint replenishment. Oper. Res. **33**(4), 820–834 (1985)
20. Chakravarty, A.K., Orlin, J.B., Rothblum, V.G.: A partitioning problem with additive objective with an application to optimal inventory groupings for joint replenishment. Oper. Res. **30**, 1018–1020 (1985)
21. Dobzinski, S., Schapira, M.: An improved approximation algorithm for combinatorial auctions with submodular bidders. In: SODA, pp. 1064–1073. ACM Press (2006)
22. Feige, U., Vondrák, J.: The submodular welfare problem with demand queries. Theory Comput. **6**(1), 247–290 (2010)
23. Ng, C.T., Barketau, M.S., Cheng, T.C.E., Kovalyov, M.Y.: "Product partition" and related problems of scheduling and systems reliability: computational complexity and approximation. Eur. J. Oper. Res. **207**(2), 601–604 (2010)

Multi-Robot Foremost Coverage
of Time-Varying Graphs

Eric Aaron[1], Danny Krizanc[2]($\boxtimes$), and Elliot Meyerson[2]

[1] Computer Science Department, Vassar College, Poughkeepsie, NY, USA
eaaron@cs.vassar.edu
[2] Department of Mathematics and Computer Science, Wesleyan University,
Middletown, CT, USA
{dkrizanc,ekmeyerson}@wesleyan.edu

Abstract. In this paper we demonstrate the application of time-varying graphs (TVGs) for modeling and analyzing multi-robot foremost coverage in dynamic environments. In particular, we consider the multi-robot, multi-depot Dynamic Map Visitation Problem (DMVP), in which a team of robots must visit a collection of critical locations as quickly as possible, in an environment that may change rapidly and unpredictably during navigation. We analyze DMVP in the context of the $\mathcal{R} \supset \mathcal{B} \supset \mathcal{P}$ TVG hierarchy. We present exact offline algorithms for k robots on edge-recurrent TVGs ($\mathcal{R}$) over a range of topologies motivated by border coverage: an $O(Tn)$ algorithm on a path and an $O(T\frac{n^2}{k})$ algorithm on a cycle (where T is a time bound that is linear in the input size), as well as polynomial and fixed parameter tractable solutions for more general notions of border coverage. We also present algorithms for the case of two robots on a tree (and outline generalizations to k robots), including an $O(n^5)$ exact algorithm for the case of edge-periodic TVGs ($\mathcal{P}$) with period 2, and a tight poly-time approximation for time-bounded edge-recurrent TVGs ($\mathcal{B}$). Finally, we present a linear-time $\frac{12\Delta}{5}$-approximation for two robots on general graphs in $\mathcal{B}$ with edge-recurrence bound Δ.

1 Introduction

For mobile robot applications such as multi-robot surveillance, search and rescue, patrol, and inspection tasks, problems are often formulated as graph coverage problems. In many such applications, the robots may navigate in dynamic environments that can change unpredictably during navigation, but conventional static graph formulations do not represent those essential dynamics. We address this issue by adopting recent formulations of *time-varying graphs* (TVGs) to enable analysis of multi-robot team navigation in dynamic environments. In particular, in this paper we present results for the multi-robot, multi-depot *Dynamic Map Visitation Problem* (DMVP), in which a team of robots must visit a collection of critical locations on a map (graph) as quickly as possible, but the environment may change during navigation. We present efficient offline algorithms, including a fixed parameter tractable solution, for an arbitrary number of robots over a

© Springer-Verlag Berlin Heidelberg 2015
J. Gao et al. (Eds.): ALGOSENSORS 2014, LNCS 8847, pp. 22–38, 2015.
DOI: 10.1007/978-3-662-46018-4_2

range of topologies motivated by border coverage (Sect. 2), and for two robots on a tree (Sect. 3); details of our main results are summarized in Sect. 1.2.

Many approaches to coverage problems [10,11] (including border coverage [14,23]) are based on static graph representations, as are related combinatorial optimization problems such as the k-traveling repairman problem, k-traveling salesman problem, etc. [4,15]. DMVP is distinct from these other problems, with crucial and distinguishing aspects of DMVP including (1) robots can start at any number of distinct depots, and (2) robots need not return to their depot after completion of coverage. Permitting multiple depots allows for the teaming of geographically disjoint robots; while completing a series of heterogeneous tasks, robots may not be together when a map visitation call is warranted. The absence of a requirement for return ensures that the singular goal is timely coverage completion, which is important for time-sensitive inspection tasks or other applications.

The most fundamental difference between DMVP and related problems is that DMVP employs a TVG representation of the environment, which can capture variation in graph structure over time in ways that static graphs cannot. A TVG [8] is a five-tuple $\mathcal{G} = (V, E, \mathcal{T}, \rho, \zeta)$, where $\mathcal{T} \subseteq \mathbb{T}$ is the *lifetime* of the system, *presence function* $\rho(e, t) = 1 \iff$ edge $e \in E$ is available at time $t \in \mathcal{T}$, and *latency function* $\zeta(e, t)$ gives the time it takes to cross e if starting at time t. The graph $G = (V, E)$ is called the *underlying graph* of $\mathcal{G}$, with $|V| = n$. As in [2,16,20], we consider the discrete case in which $\mathbb{T} = \mathbb{N}$, edges are undirected, and all edges have uniform travel cost 1. If agent a is at u, and edge (u, v) is available at time τ, then a can take (u, v) during this time step, visiting v at time $\tau + 1$. As a traverses $\mathcal{G}$ we say a both *visits* and *covers* the vertices in its traversal, and we will henceforth use these terms interchangeably. $\mathcal{J} = \{(e_1, t_1), ..., (e_k, t_k)\}$ is a *journey* $\iff \{e_1, ..., e_k\}$ is a walk in G (called the *underlying walk* of $\mathcal{J}$), $\rho(e_i, t_i) = 1$ and $t_{i+1} \geq t_i + \zeta(e_i, t_i)$ for all $i < k$. The *topological length* of $\mathcal{J}$ is k, the number of edges traversed. The *temporal length* is the duration of the journey: (*arrival date*) − (*departure date*). Given a date t, a journey from u to v departing on or after t whose arrival time is soonest is called *foremost*; whose topological length is minimal is called *shortest*; and whose temporal length is minimal is called *fastest*.

In [8], a hierarchy of thirteen TVG classes is presented. In related work on exploration [2,16,19], broadcast [7], and offline computation of optimal journeys [6], focus is primarily on the chain $\mathcal{R} \supset \mathcal{B} \supset \mathcal{P}$, which enforce natural constraints for mobile robot applications: $\mathcal{R}$ (recurrence of edges) is the class of all TVG's $\mathcal{G}$ such that G is connected, and $\forall e \in E, \forall t \in \mathcal{T}, \exists t' > t$ s.t. $\rho(e, t') = 1$; $\mathcal{B}$ (time-bounded recurrence of edges) is the class of all TVG's $\mathcal{G}$ such that G is connected, and $\forall e \in E, \forall t \in \mathcal{T}, \exists t' \in [t, t + \Delta)$ s.t. $\rho(e, t') = 1$, for some Δ; $\mathcal{P}$ (periodic edges) is the class of all TVG's $\mathcal{G}$ such that G is connected, and $\forall e \in E, \forall t \in \mathcal{T}, \forall k \in \mathbb{N}, \rho(e, t) = \rho(e, t + kp)$ for some p, the *period* of $\mathcal{G}$.

We are interested in solving the following problem:

Problem. *Given a TVG $\mathcal{G}$ (in class $\mathcal{R}$, $\mathcal{B}$ or $\mathcal{P}$) and a set of starting locations S for k agents in $\mathcal{G}$, the TVG foremost coverage or Dynamic Map Visitation*

Problem (DMVP) is the task of finding journeys starting at time 0 for each of these k agents such that every node in V is in some journey, and the maximum temporal length among all k journeys is minimized. The decision variant asks whether this coverage can be completed in no more than a given t time steps, that is, these journeys can be found such that no journey has arrival date later than t.

For the minimization version of the problem $DMVP(\mathcal{G}, S)$ and the corresponding decision problem $DMVP(\mathcal{G}, S, t)$, the input is viewed as a sequence of graphs G_i each represented as an $n \times n$ adjacency matrix, with an associated integer duration t_i, i.e., $\mathcal{G} = (G_1, t_1), (G_2, t_2), ..., (G_m, t_m)$, where G_1 appears initially at time zero (see [9, 21, 22] for alternative views of TVGs). Let $T = \sum_{i=1}^{m} t_i$. We know from [2] that we can run an $O(nm)$ preprocessing step that lets us presume that $T < 2\,nm\text{--}3\,m$, (that is, T is at worst linear in $\mathcal{G}$), and enables $O(1)$ edge presence lookups $\rho(e, \tau)$, without affecting asymptotic runtime of any of the algorithms presented below. We think of the input $\mathcal{G}$ as a temporal subgraph of some TVG $\mathcal{G}_\infty$ with lifetime $\mathbb{N}$ and the same edge constraints as $\mathcal{G}$. Thus, the limited information provided in $\mathcal{G}$ is used to find journeys (which may have temporal length greater than T) that cover G, for agents in $\mathcal{G}_\infty$.

1.1 Related Results

The problem most similar to (but distinct from) DMVP is the minmax k-traveling salesman problem [12, 27], in which all robots start at and return to a single depot on a static graph. Approximation algorithms have been given that forgo the single depot requirement, but still require a return to multiple initial depots [4, 26]. To the best of our knowledge, no previous work has addressed the case of exact algorithms for multiple agents either without return or with multiple depots, even for the static case. A pseudo-polynomial time algorithm for any constant $k > 1$ agents on a tree for the k-traveling salesman problem (single depot with return) is presented in [12]. A pseudo-polynomial solution for the weighted tree case is given in [27]. This algorithm runs in $O(n^3)$ for the restriction to two robots and unweighted edges (Lemma 1). We sequentially generalize this to DMVP by (1) allowing multiple depots, (2) not requiring robots to return to their depots, and (3) incorporating TVG models, namely, $\mathcal{P}$ and $\mathcal{B}$ (Sect. 3).

Heuristics for boundary coverage for multiple robots are considered in [14], in which the problem is reduced to k-rural postman over a static graph extracted from a continuous environment. This graph extraction procedure motivates our result on "border coverage" graphs in $\mathcal{R}$ (Theorem 3).

The complexity of DMVP for a single agent was explored in [2], in which it was shown that in the edge-recurrent TVG class $\mathcal{R}$ it is NP-hard to approximate within any factor, even over max-degree 3 trees, and stars (i.e., trees with at most one vertex of degree greater than 1). (A related result was derived independently in [25].) The periodic case $\mathcal{P}$, even with period $p = 2$, was shown to be NP-hard over a larger class of graphs than the static case MVP, which is hard even over trees when k is part of the input [1]. Other hardness results for problems over TVGs have been shown for computing strongly connected components [5] and dynamic diameter [18].

1.2 Main Results

We present algorithms for DMVP for k agents in $\mathcal{R}$ over a range of topologies motivated by border coverage: an $O(Tn)$ algorithm to optimally solve DMVP on a path, an $O(T\frac{n^2}{k})$ algorithm on a cycle, a polynomial solution for the border graph of a planar region divided into a constant number of components, and a fixed parameter tractable solution for any m-leaf c-almost tree, for parameters m and k, and constant c. We demonstrate a fundamental hardness separation between $\mathcal{P}$ and static graphs for all fixed k. We then consider the case of trees in $\mathcal{P}$ with $p = 2$ and give a $O(n^5)$ algorithm for the case of two agents. We also give an $O(n^3)$ algorithm for tight approximation for two agents on a tree in $\mathcal{B}$. Finally, we present a linear-time $\frac{12\Delta}{5}$ approximation for two agents on general graphs in $\mathcal{B}$ with edge-recurrence bound Δ. Corresponding generalizations to k agents are outlined here and will appear in the full version of this paper.

2 k-Agent Border Coverage in $\mathcal{R}$

DMVP on paths, cycles and more general classes of graphs is motivated by border coverage, e.g., for security. Coverage of a path corresponds to securing critical points along the border between any two adjacent connected planar regions, neither of which surrounds the other, while coverage of a cycle corresponds to securing the complete border of any simply connected planar region.

Theorem 1. *DMVP for k agents in $\mathcal{R}$ on a path is solvable in $O(Tn)$ time.*

Proof. Consider DMVP with underlying graph the path $P = v_1...v_n$ and k agents $a_1, ..., a_k$ starting at locations $s_1, ..., s_k$, respectively. Orient P left-to-right, with v_1 the leftmost vertex. Without loss of generality, suppose $s_1, ..., s_k$ are ordered from left to right. Note that if two or more agents start at the same vertex s_i, simply sending two of them in opposite directions will be trivially optimal, thereby reducing the problem to two instances of DMVP over edge-disjoint subpaths $v_1...s_i$ and $s_i...v_n$, which can be solved independently.

Assume no two agents start at the same node. The idea is to compute for each vertex $u \in s_1...v_n$ the optimal cost of the solution to the DMVP subproblem over $v_1...u$ for all agents starting on or to the left of u. Call this cost $c(u)$. We can compute all $c(u)$ from left to right, and finally get the result $c(v_n)$ for DMVP for all k agents over P (Algorithm 1). On a path, it is never advantageous for any two agents to cross over one another, since they could simply each turn around instead. As a result, agent a_1 must cover v_1. Let v be the node directly to the left of s_2. The subproblems to be computed from $c(s_1)$ to $c(v)$ concern only agent a_1. $c(s_1)$ is the time it takes a_1 to reach v_1 by simply traveling left starting at

Algorithm 1. DMVP-Path($\mathcal{G}, \{s_1, ..., s_k\}$)

---

for all $v \in s_1...v_n$ **do**   ▷ Initialize c
  $c(v) = \infty$
for $i = 1, ..., k$ **do**
  $lBoundary = s_i$ ▷ Evaluate all left-first journeys for a_i
  if $i = 1$ **then**
    $lBoundary = v_1$
  while $lBoundary \notin \{s_{i-1}, \emptyset\}$ **do**   ▷ Try every possible left endpoint
    $t = 0$
    $loc = s_i$
    $turned = eval = False$ ▷ evaluate solution?
    while $loc \notin \{\emptyset, s_{i+1}\}$ and $t < T$ **do**   ▷ enter at most T times
      if $loc = lBoundary$ **then**
        $turned = True$
      if $turned = True$ and $loc = s_i$ **then**
        $eval = True$
      if $eval = True$ **then**
        $c(loc) = \min(c(loc), \max(c(lBoundary.lNode), t))$
      if not $turned$ and $\rho(loc.lEdge, t) = 1$ **then**
        $loc = loc.lNode$
      if $turned$ and $\rho(loc.rEdge, t) = 1$ **then**
        $loc = loc.rNode$
      $t = t + 1$
    $lBoundary = lBoundary.lNode$
  $rBoundary = s_i$ ▷ Evaluate all right-first journeys for a_i
  if $i = k$ **then**
    $rBoundary = v_n$
  while $rBoundary \notin \{s_{i+1}, \emptyset\}$ **do**   ▷ Try every possible right endpoint
    $t = 0$
    $loc = s_i$
    $turned = eval = False$ ▷ evaluate solution?
    while $loc \notin \{s_{i-1}, \emptyset\}$ and $t < T$ **do**   ▷ enter at most T times
      if $loc = rBoundary$ **then**
        $turned = True$
      if $turned = True$ and $loc = s_i$ **then**
        $eval = True$
      if $eval = True$ **then**
        $c(rBoundary) = \min(c(rBoundary), \max(c(loc.lNode), t))$
      if not $turned$ and $\rho(loc.rEdge, t) = 1$ **then**
        $loc = loc.rNode$
      if $turned$ and $\rho(loc.lEdge, t) = 1$ **then**
        $loc = loc.lNode$
      $t = t + 1$
    $rBoundary = rBoundary.rNode$
  return $c(v_n)$

---

time 0. For all u strictly between s_1 and s_2, a_1 can cover $v_1...u$ either by going left first or right first. We can compute all left-first journeys in a single pass in $O(T)$ by going left until hitting v_1, then turning around and recording the time at which each u is reached. For the journeys that go right first, a_1 travels right to u, turns around and travels left until v_1 is reached. For each u, the minimum of the left-first and right-first journey is stored as $c(u)$. Doing this for each u takes overall $O(T|s_1...s_2|)$.

Now consider any agent a_i in $\{a_2, ..., a_{k-1}\}$, and suppose all subproblems to the left of s_i have already been computed. Let L_i be the path from the right neighbor of s_{i-1} to s_i, and R_i be the path from s_i to the left neighbor of s_{i+1}. In a full optimal solution over P, the leftmost vertex a_i covers could be any vertex in L_i, and the rightmost vertex could be any in R_i. $c(s_i)$ is the minimum over all v_j in L_i, of the maximum of $c(v_{j-1})$ and the time it takes a_i to reach v_j traveling left from time 0. This is computed in a single $O(T)$ left pass. Now suppose the rightmost vertex a_i covers is not s_i. Then, if a_i goes left first and turns around at l, we can compute the cost of a_i's journey ending at each vertex $r \neq s_i$ in R_i in a single $O(T)$ pass, in which a_i turns around at l and then travels right as far as possible. Doing this for each l takes overall $O(T|L_i|)$. Similarly, if a_i goes right first and turns around at r, we can compute the cost of a_i's journey ending at each vertex $l \neq s_i$ in L_i in a single $O(T)$ pass, in which a_i turns around at r and then travels left as far as possible. Doing this for each r takes overall $O(T|R_i|)$. For each $r \in R_i$, $c(r)$ is the minimum over all $v_j \in L_i$, of the maximum of $c(v_{j-1})$ and the minimum between the left-first and right-first journeys of a_i covering $v_j...r$. $c(r)$ can simply be updated immediately anytime a better solution is evaluated.

a_k faces a similar situation to a_1, it must cover v_n, so only needs to consider variable left endpoints. The cost of the optimal solution over all of P is then the minimum over all $v_j \in L_k$ of the max of $c(v_{j-1})$ and the minimum between the left-first and right-first journeys of a_k covering $v_j...v_n$. Computation of the complete DMVP solution over P takes $O(T|R_1|) + O(T|L_2|) + O(T|R_2|) + ... + O(T|L_{k-1}|) + O(T|R_{k-1}|) + O(T|L_k|) = O(Tn)$. □

Theorem 2. *DMVP for k agents in $\mathcal{R}$ on a cycle is solvable in $O(T\frac{n^2}{k})$ time.*

Proof. Consider DMVP over the cycle $C = v_0v_1...v_nv_0$ for k agents $a_1, ..., a_k$ ordered clockwise around the cycle at locations $s_1, ..., s_k$, respectively. If any two agents start at the same node, then sending them in opposite directions will be optimal, thereby reducing the problem to DMVP on a path, which can be solved with Algorithm 1 in $O(Tn)$. If no two agents start at the same node, let d be the shortest distance between any two agents a_i, a_{i+1}. Since there are k agents, $d \leq \lfloor \frac{n}{k} \rfloor$. The furthest that a_{i+1} covers counter-clockwise can be any node from s_{i+1} to the immediate clockwise neighbor of s_i. For each of these $O(n/k)$ potential left endpoints v_j, we can run Algorithm 1 on the path consisting of C with the edge (v_{j-1}, v_j) removed. Taking the minimum over all v_j results in an $O(T\frac{n^2}{k})$ runtime. □

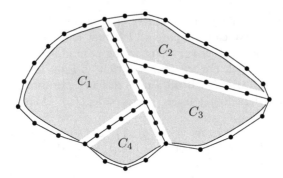

Fig. 1. Border coverage graph extracted from a planar region (gray) subdivided into four components.

The next result corresponds to a more general notion of border coverage akin to that addressed in [14]. Consider any simply connected planar region divided by throughways into some number of subregions, e.g., a complex of secure buildings or zones. The *border coverage graph* of such a subdivided region is the graph induced by the coverage of critical points along the union of the subregions' borders, e.g., Fig. 1.

Theorem 3. *DMVP for k agents in $\mathcal{R}$ is solvable in $O(Tn^{6c+1})$ time, when G is the border coverage graph of a simply connected planar region divided into c subregions, for c constant.*

Proof. Suppose R is a simply connected planar region divided into c subregions, for c constant. Call a path $P \subset G$ a *through-path* if the endpoints of P have degree greater than two and all intermediate vertices of P have degree two. Let c_1 be the number of through-paths in G. c_1 can be bounded by considering how R is subdivided. Let G_i be the border coverage graph corresponding to R divided into $i < c$ subregions. We create a new subregion by adding a through-path P between two vertices of G_i, such that all vertices of P are internal to R and no edge of P crosses an edge of G_i. This addition creates another through-path for each endpoint of P that had degree two before the addition. Thus, at most three new through-paths are added for each subregion of R, i.e., $c_1 < 3c$.

In an optimal solution, the agents that start on a through-path $P = u...v$ but never reach u or v must together cover a set of vertices whose induced graph is a subpath of P. For each P, there are $O(n^2)$ such subpaths. It would never be better for an outside agent to enter P in order to cover a vertex between two disjoint subpaths, as it must cross over at least one agent that never leaves P, and the remainder of their journeys could be swapped at no cost. So we forbid outside agents to travel along these subpaths. From Theorem 1, DMVP for each of these subpaths can be computed in $O(Tn)$.

Selecting the subpath these agents cover for every through-path induces a subset of the remaining vertices that must be covered to complete coverage, namely, the vertices adjacent to but not included in any subpath. There are $O(n^{2c_1})$ ways to make this selection. If the internally-covered subpath of a through-path P with endpoints u and v is empty, then it must be that no agents started between u and v, so in addition to the $O(n)$ choices for pairs of vertices adjacent to a subpath of P of length 0, there are two further ways for outside agents to complete coverage of P: by at some point traveling directly from u to v, or from v to u, covering all of P along the way. At most two outside agents are required to cover the remainder of each path, so in an optimal solution at most $2c_1$ agents leave their start paths. For any P, the agents that could leave are each of the at most $2c_1$ closest to u and v, resp. There are $c_2 = (2c_1)^{4c_1^2}$ ways to partition the remaining elements to cover between all agents that could leave their start paths, and after running an $O(Tn^3)$ all-pairs-all-times-foremost-journey preprocessing step [2], DMVP for each agent can be computed in $O(4c_1^2 2^{2c_1})$. Running this for each agent for the $O(n^{2c_1})$ ways to cover all paths and computing internal path costs yields a total runtime of $O(n^{2c_1})(O(4c_2c_1^2 2^{2c_1}) + O(c_1 Tn)) + O(Tn^3) = O(Tn^{6c+1})$.   □

Pointing towards further generalizations, the following theorem extending Thm. 10 in [2] applies to a slightly larger classes of graphs and includes the number of agents as a parameter in an fixed parameter tractable (FPT) solution. We will give the complete proof in the full version.

Theorem 4. *DMVP for k agents in $\mathcal{R}$ is fixed parameter tractable, when G is an m-leaf c-almost-tree, for parameters m and k, and c constant.*

Proof sketch. For all $t < T$, consider the decision variant over m-leaf trees. Partitioning the leaf set among agents (k^m ways) and using the single agent $O(Tn^{3+c}f(m))$ algorithm [2] for each guarantees coverage of everything but the union of shortest paths between depots. If an edge in such a path has not already been covered this creates a cut with agents confined to subtrees. There are 2^k ways to select which paths are cut. Such a selection induces a tree of subproblems which can be solved in a bottom-up fashion, fixing cut points along the way. There will always be an unsolved subproblem of degree no more than one. Fix the cut point for this subproblem as far as possible given the time bound t by testing the instance corresponding to each of the $O(n)$ possible cut points along the path. This factor of n can be pulled out to keep the algorithm FPT for k, m. It is straightforward to extend this idea to c-almost-trees.   □

3   Two Agents on a Tree

We know from [2] that DMVP for k agents on a tree is hard in $\mathcal{B}$, regardless of k, even over spiders (i.e., trees in which at most one vertex has degree greater than 2). However, for a single agent in $\mathcal{P}$, DMVP can be solved in polynomial time over spiders for fixed p, and in linear time on arbitrary trees when $p = 2$. Since in $\mathcal{P}$ we are able to efficiently solve DMVP over a wider range of graph classes than in $\mathcal{B}$ or $\mathcal{R}$ [2], to show that for multiple agents $\mathcal{P}$ is fundamentally more complex than the class of static graphs, we demonstrate that for DMVP

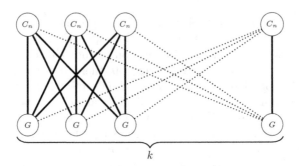

Fig. 2. Graph class for which DMVP is NP-hard in $\mathcal{P}$ with $p = 2$, but trivially in P when $p = 1$. Each thick edge represents the edges in the complete biparitite graph linking vertices in a C_n to vertices in a G.

for k agents there is a hardness separation between $\mathcal{P}$ with $p = 2$, and $p = 1$, for all k. Note that when $p = 2$, edges can only be one of three possible dynamic types: (01) available only at odd times, (10) available only at even times, (11) available at all times.

Theorem 5. *For all $k \geq 1$, there is a class of graphs C such that DMVP in $\mathcal{P}$ for k agents over graphs in C is trivial when $p = 1$, but NP-hard when $p = 2$.*

Proof. For any graph G with an even number of vertices $v_0, ..., v_{n-1}$, take k copies of G and k copies of $C_n = c_0...c_{n-1}c_0$, a cycle of length n. Add edges to form a complete bipartite graph linking vertices in each C_n to each G (see Fig. 2). For $p = 2$, let all original edges of G be of type 11. Let all (v_i, c_i) be of type 01 when i is even and type 10 when i is odd. Let (v_i, c_{i+1}) and (c_i, c_{i+1}) be of type 10 when i is even and 01 when i is odd, where indices are taken mod n. Suppose each agent a_i starts at a distinct v_0. If $t = 2n - 1$, a_i must completely cover G before moving to a C_n, to avoid waiting at a vertex of C_n for a time step on the way back to a G, and thus effectively solve HAM-PATH [17] on G. However, when $p = 1$, each agent simply jumps repeatedly from a G to a C_n, since every uncovered vertex across the bipartite cut is always available.   □

Now, even DMVP restricted to static graphs (also known as MVP) is in general NP-hard on trees for k agents, but for a single agent it can be solved in linear time [1]. What about DMVP when $k = 2$? We build up to an exact polynomial solution for DMVP on a tree for two agents a_1, a_2 in $\mathcal{P}$ for the $p = 2$ case, and a tight approximation in $\mathcal{B}$ for all Δ, via a series of related lemmas partially-ordered by constraints, see Fig. 3. The base result (Lemma 1) is implied by an upper bound established in [27], but the further results are, to our knowledge, novel generalizations, with our main result being an $O(n^5)$ solution for DMVP in $\mathcal{P}$ for two agents with $p = 2$ (Theorem 7).

Lemma 1. *MVP with return for two agents starting at a single depot on a tree can be solved in $O(n^3)$ time.*

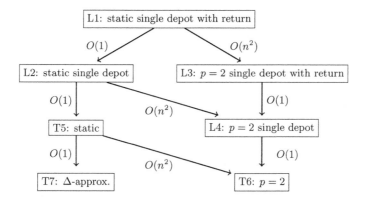

Fig. 3. Poset of results leading to solutions for two-agent DMVP on a tree; arrows indicate increasing factors of complexity as constraints are loosened.

Proof sketch. This result is implied as a special case in [27]. (An $O(n^6)$ algorithm is given in [12].) We give the following proof idea:

At each node v (whose maximal subtree is denoted G^v) from the leaves up to the starting depot, i.e., root s, we compute and store possible pairs of costs for a_1 and a_2 covering and returning to the root of the maximal subtree rooted at v, by iterating over the pairs of costs $(c_1, c_2)_{u_i}$ associated with covering each of v's children $u_1, ..., u_{\deg v}$, to compute partial solution costs $(c_1, c_2)_v^i$, corresponding to possible pairs of costs covering the subtrees $G^{u_1}, ..., G^{u_i}$. However, each new cost pair $(c_1, c_2)_v^i$ is only stored if c_2 is less than the current best cost associated with c_1. Each cost is bounded by $2n - 3$, and each child is iterated over only once, taking $O(n^2)$ to combine its costs $(c_1, c_2)_{u_i}$ with the costs $(c_1, c_2)_v^{i-1}$ of covering the previous branches of the subtree to get all $(c_1, c_2)_v^i$, yielding the $O(n^3)$ total runtime. □

The following extension drops the constraint of returning to root.

Lemma 2. *MVP for two agents starting at a single depot on a tree can be solved in $O(n^3)$ time.*

Proof. Follow the same method describe in the proof of Lemma 1, except now at each node v store pairs of costs for covering G^v for each of the following four cases: both a_1 and a_2 return to v $((c_1^r, c_2^r)_v)$, a_1 returns to v but not a_2 $((c_1^r, c_2)_v)$, a_2 returns to v but not a_1 $((c_1, c_2^r)_v)$, neither returns to v $((c_1, c_2)_v)$. Partial solutions are then updated for each return type: $(c_1^r, c_2^r)_v^{i-1}$ combined with $(c_1^r, c_2^r)_{u_i}$, $(c_1^r, c_2)_{u_i}$, $(c_1, c_2^r)_{u_i}$, and $(c_1, c_2)_{u_i}$ to get $(c_1^r, c_2^r)_v^i$, $(c_1^r, c_2)_v^i$, $(c_1, c_2^r)_v^i$, and $(c_1, c_2)_v^i$, resp.; $(c_1^r, c_2)_v^{i-1}$ combined with $(c_1^r, c_2)_{u_i}$ and $(c_1, c_2)_{u_i}$ to get $(c_1^r, c_2)_v^i$ and $(c_1, c_2)_v^i$; $(c_1, c_2^r)_v^{i-1}$ combined with $(c_1^r, c_2^r)_{u_i}$ and $(c_1^r, c_2)_{u_i}$ to get $(c_1, c_2^r)_v^i$ and $(c_1, c_2)_v^i$; $(c_1, c_2)_v^{i-1}$ combined with $(c_1^r, c_2^r)_{u_i}$ to get $(c_1, c_2)_v^i$. That is, with-return costs are added to with-return costs to get new with-return costs, as the journey must end on some later branch; without-return costs are added to with-return costs to get new without-return costs that end on the current branch; with-return costs are added to without-return costs to get new without-return costs that end up on some previous branch. Updating cost pairs for all four return types incurs only a constant factor runtime increase over the return to root case.   □

This is generalized now to the case of multiple depots; the standard MVP formulation.

Theorem 6. *MVP for two agents on a tree can be solved in $O(n^3)$ time.*

Proof. Suppose a_1 starts at s_1 and a_2 starts at s_2. Let $P = (s_1 = p_1)p_2...p_{l-1}$ $(p_l = s_2)$ be the unique simple path from s_1 to s_2.

First, note that if a_1 and a_2 do not cross paths, that is, there is no $v \in P$ such that both a_1 and a_2 include v in their journeys, then the subtrees covered by a_1 and a_2 will be disjoint, reducing the problem to two instances of MVP on a tree for a single agent, each of which can be solved independently in $O(n)$ [1]. There are $O(n)$ ways to cut P so that the journeys are disjoint, so trying each of these possibilities takes $O(n^2)$, which is subsumed by the cost of considering non-disjoint solutions.

Assuming the optimal journeys are not disjoint, using the algorithm described in Lemma 2, run for all $v \in P$ MVP for two agents starting at a single depot for the maximal subtree rooted at v that is edge-disjoint from P, generating all resulting potential cost pairs. Now, we build up solutions from left-to-right, i.e., from s_1 to s_2. After considering each p_i along P, we want all cost pairs for all four cost pair cases (both return, only a_1 returns, etc.) of covering all of G excluding the branches rooted at all p_j, for all $j > i$. With-return costs are added to with-return costs to get new with-return costs; without-return costs are added to with-return costs to get new without-return costs that end on p_i's branch; with-return costs are added to without-return costs to get new without-return costs that end up on some previous branch rooted at p_k, for some $k < i$. Additional cost for traversing P is accumulated along the way: each time a_1 precedes to the next vertex of P, 1 is added to the cost of a_1's with-return costs (2 to without-return); $2|l - i|$ added to a_2's costs when p_i is selected to be its furthest vertex reached, and $|l - j|$ subtracted when p_j's branch is marked as the final branch a_2 enters, i.e., when p_j's without-return costs are added to previous with-return costs for $p_1...p_{j-1}$. Updating costs at each branch again takes $O(n^3)$, so the cost of the overall solution remains $O(n^3)$.   □

That concludes our results for the static case. We now generalize these results to the case of TVGs in $\mathcal{P}$.

Lemma 3. *DMVP with return for two agents starting at a single depot on a tree in $\mathcal{P}$ can be solved in $O(n^5)$, when $p = 2$.*

Proof. This case runs similar to Lemma 1, but since we are in $\mathcal{P}$, we must be careful about how we build up solutions, as it matters in which order branches are taken. From [2], we know that with $p = 2$, each agent can enter each branch at most once, and that each branch can in an $O(n)$ pre-processing step be marked as either 01, fastest journey available only at even times; 10, fastest journey only available at odd times; or 11, fastest journey always available. Note that this also applies to the subbranch covered by each agent. Furthermore, the optimal way for a single agent to cover any set of classified branches is to alternate between taking 01's and 10's as many times as possible, before taking the remaining branches in any order. So, given a start time along with the difference d_i between the number of 01's and 10's in a partial solution for a_i's coverage of G^v, we can add a partition of a new branch and check in constant time exactly how the cost of our solution will be affected.

Computing from the leaves up, as in Lemma 1, we store all possible pairs of costs of covering the maximal subtree rooted at v, but now we store separate pairs of costs for four cases defined by whether each a_i reaches v at an odd or even time. This adds only a constant factor overhead, and given a pair of costs and start times, we can in constant time compute whether the type τ_i of each of the two journeys is 01, 10, or 11. Storing all possible d_i for each cost pair means $O(\deg(v)^2 n)$ tuples (c_1, c_2, d_1, d_2) are stored at each branch, with branch updates taking $O(\deg(v)^2 n^2)$, as $(c_1, c_2, d_1, d_2)_v^{i-1}$ partial solutions are combined with $(c_1, c_2, \tau_1, \tau_2)_{u_i}$, yielding a cost of $O(\deg(v)^3 n^2)$ per node, and hence $O(n^5)$ overall.   □

Lemma 4. *DMVP for two agents starting at a single depot on a tree in $\mathcal{P}$ can be solved in $O(n^5)$ time, when $p = 2$.*

Proof. In a similar manner to the extension from Lemma 1 to Lemma 2, we now store cost pairs for each of the four return cases for a_1 and a_2. Branch types can still be maintained as in Lemma 3 in order to preserve optimal orderings, so the algorithm again runs in $O(n^5)$.   □

Theorem 7. *DMVP for 2 agents on a tree in $\mathcal{P}$ can be solved in $O(n^5)$ time, when $p = 2$.*

Proof. Again, let $P = (s_1 = p_1)p_2...p_{l-1}(p_l = s_2)$ be the unique simple path from s_1 to s_2, and assume a_1's and a_2's journeys are not disjoint, since we can solve each of the $O(n)$ disjoint instances in $O(n)$. We adopt a similar though more involved version of the left-to-right dynamic programming approach from the proof of Lemma 6, as the order now matters in which each agent takes its portion of each subtree b_i (i.e., subtrees rooted at each p_i but disjoint from P). First, compute all pairs of costs for covering each b_i in $O(n^5)$ via Lemma 4.

Suppose the final subtree taken by a_1 in an optimal solution is b_j. Then, a_1 can only take its assigned sections of $b_1, ..., b_{j-1}$ as it moves towards s_2 for the first time, since each agent can enter any subtree at most once in $\mathcal{P}$ with $p = 2$. Suppose the closest a_1 gets to s_2 is p_k. Then all of $b_{j+1}, ..., b_{k-1}$ can be taken either on the way from p_j to p_k, or on the way back. A similar case applies for a_2. So, as we consider each branch from s_1 to s_2, we build up partial solution costs for a_1 and a_2 in two directions at once: outside-in for a_1, and inside-out for a_2. That is, suppose that through $i - 1$ branches we have stored all $(c_1(\rightarrow), c_1(\leftarrow), t)_P^{i-1}$, where $c_1(\rightarrow)$ is the cost of the forward journey so far, $c_1(\leftarrow)$ is the cost of the reverse journey (i.e., after covering its portion of b_k),

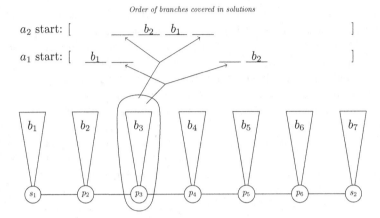

Fig. 4. Possible ways to update costs for a pair of partial solutions to include each agent's coverage of b_3, assuming a_1 ends on b_2, a_2 ends on some branch b_i, with $i > 3$, and a_2 takes b_2 on the way to b_1. In this case, both a_1 and a_2 can take their portion of b_3 either directly after b_1 or directly preceding b_2.

and t is the start time for the reverse journey; and $(c_2, t)_P^{i-1}$, where c_2 is the costs of a_2 covering its portions of $b_1, ..., b_{i-1}$, assuming p_{i-1} is reached by a_2 for the first time at time t. Update partial solutions to include b_i in the following way: for a_1, if a branch has been taken without return, further branches can be taken either directly after all forward journeys or directly before all reverse journeys, otherwise, all branches can only be taken on forward journeys, except of course for the branch taken without return, which must be taken last; for a_2, if a branch has been taken without return, further branches must be taken with return directly preceding existing solutions, otherwise, further branches can either be taken directly before existing solutions *or* directly after (see Fig. 4), and all must be taken with return. The cost of the final branch taken by a_1 is succinctly inserted between the forward and backward costs. Additional costs accumulated via the traversal of edges of P are added in as in Theorem 6, but now taking into account the edge type (i.e., 01,10, or 11), and the time parity at which the edge is reached. Running these updates for each time parity, each return case, and each location of the branch in an optimal ordering incurs together only constant factor overhead. Storing both $(c_2, t)_P^i$ for all $(c_1(\rightarrow), c_1(\leftarrow), t)_P^i$, takes $O(n^2)$ space, but we can reduce this to $O(n)$ by compactly representing the solution cost by the sum of $c_1(\rightarrow)$ and $c_1(\leftarrow)$, and a bit for storing the parity of each. The update at each branch still takes $O(n^2)$ to compute all possible cost pair cases for the new partial solutions, so the full iteration from p_1 to p_l takes $O(n^3)$, and the initial $O(n^5)$ runtime dominates. $\square$

We can also apply Theorem 6 to get a tight approximation for two agents on a tree in $\mathcal{B}$:

Theorem 8. *DMVP for two agents on a tree in $\mathcal{B}$ can be Δ-approximated in $O(n^3)$ time $\forall \Delta > 1$. This approximation is tight.*

Proof. The cost to cover a tree G in $\mathcal{B}$ for two agents, starting at potentially distinct depots, is lower-bounded by the cost C of covering the static G from these same depots. C can be computed in $O(n^3)$ via the algorithm described in Theorem 6. By following in $\mathcal{B}$ the journeys resulting in static cost C, each agent will wait at most $\Delta - 1$ steps for each successive edge to appear, thereby completing coverage in no more than ΔC steps. Since C is the fastest possible cost of covering G, this must be a Δ-approximation.

It is straightforward to extend to the case of k agents the result from [2] that DMVP for a single agent in $\mathcal{B}$ over trees is NP-hard to approximate within any factor less than Δ; simply link together by long paths k copies of the graph constructed for that proof.   □

Over general graphs in $\mathcal{B}$, we can use spanning tree coverage to get the following approximation:

Theorem 9. *DMVP for two agents in $\mathcal{B}$ can be $\frac{12\Delta}{5}$-approximated in $O(n)$ time* $\forall \, \Delta > 1$.

Proof. Given a graph G, and a spanning tree H of G (constructed in $O(n)$ time), the Euler tour of H is a $2n - 1$ node cycle C, the complete coverage of which implies complete coverage of G. From [1], for a cycle, we know each agent covers no more than $\lceil \frac{3}{5} \rceil |C| - 2$ edges in an optimal two agent solution, which can be found in $O(n)$ time. Following this solution in $\mathcal{B}$, the two agents take at most $\Delta(\lceil \frac{6n-3}{5} \rceil - 2)$ steps to complete coverage, which is no more than $\frac{12\Delta}{5}$ times worse than the minimum possible number of steps $\lceil \frac{n-1}{2} \rceil$ for covering G.   □

We are able to extend Theorems 6 and 7 to any fixed number of agents k, applying ideas from the extension of 2-partition to k-partition for multisets of integers. With multiple depots, the union of the shortest paths between depots forms a k-leaf tree H. The possible costs of partitioning subtrees rooted at vertices in H but edge-disjoint from H and covering these subtrees along a path between two depots can be computed in a similar manner to the proof of Theorem 7. Then, the method of optimally ordering 01, 10, and 11 branches can be used on H itself. The methods for establishing approximation bounds for Theorem 8 will also still hold in the k agent case. For Theorem 9, bounds for cycle coverage in [1] enable $k\Delta$-approximations in $O(kn^3)$ for any $k > 2$. We will give the complete proofs in the full version of this paper.

4   Conclusion and Discussion

This paper has demonstrated the use of time-varying graphs for modeling multi-robot foremost coverage in dynamic environments, through consideration of the Dynamic Map Visitation Problem (DMVP). We have presented efficient algorithms for an arbitrary fixed number of agents for a range of topologies motivated by border coverage, and for two agents on a tree. Future work will extend Theorems 6, 7, 8 and 9 to a polynomial time solution for any fixed k, and we believe it is also possible to make the extension to fixed p in Theorem 7. This begins by extending the idea that "when $p = 2$, an agent can enter any subtree at most once" to "for any $p > 1$, an agent at o can visit a node at depth $p - 1$ in G^o and return to o at most once".

In general, allowing for the number of agents to not be fixed increases the complexity of the problem, but when the number of agents becomes linear in the size of the graph—or, in the case of trees, linear in the number of leaves—special behavior can occur that further exposes the implications of applying constraints on edge dynamics and the number of depots to this type of problem. We make the following observation:

Remark 1. DMVP for $\frac{n}{c}$ agents on an star can be solved in polynomial time in $\mathcal{B}$ for any fixed Δ, but is hard in $\mathcal{R}$, for c constant.

Proof. Recall that a star is a tree in which at most one vertex has degree greater than 1. In $\mathcal{B}$, DMVP is upper-bounded by $2\Delta c$, when each agent is assigned c vertices to cover, and the foremost journeys taken between each are as long as possible. At each time step, each agent has $O(n)$ ways to continue its journey, so computing all possible journeys of time $\geq 2\Delta c$ for all agents takes $O((\frac{n^2}{c})^{2\Delta c}) = O(n^{4\Delta c})$. In $\mathcal{R}$, we cannot upper-bound the length of solutions, so all agents except one may be trapped together at a single vertex indefinitely, while the remaining agent is left alone to cover the rest of the star itself, which is NP-hard [2]. □

More generally, DMVP becomes tractable whenever it is possible to upper-bound optimal solutions by some constant, e.g., in $\mathcal{B}$ and $\mathcal{P}$ as k approaches n. This idea complements results of fixed parameter tractability for problems over TVGs of fixed treewidth [24], in which T is a fixed parameter.

Remark 2. With a single depot, DMVP for $k \geq m$ agents on an m-leaf tree is easy in $\mathcal{R}$; but with two depots, it is hard in $\mathcal{B}$, for all Δ.

Proof. If all agents start at a single depot s, sending one agent to each leaf l via the foremost journey from s to l will be optimal, even in $\mathcal{R}$. Now, consider the situation in $\mathcal{B}$, with $\Delta = 2$ over a spider with one sufficiently long leg. If one agent a starts at the center of the spider and the rest at the end of the long leg, in an optimal solution, a must cover all of the other legs before any other agent reaches the center, so the problem reduces to a single agent on a spider, which we know is hard. □

Decisive factors for DMVP tractability include environment topology, number of robots, and also the number of depots. The challenges of intractability that arise from these generalizations motivate research into online solutions to the problem. As a related example, [13] includes online approaches to static tree exploration with limited communication between agents. In future work, we plan to extend our results to markovian TVG models (e.g., [3,9]), which could support online solutions for general cases of map visitation problems in probabilistic dynamic environments.

References

1. Aaron, E., Kranakis, E., Krizanc, D.: On the complexity of the multi-robot, multi-depot map visitation problem. In: IEEE MASS, pp. 795–800 (2011)
2. Aaron, E., Krizanc, D., Meyerson, E.: DMVP: Foremost waypoint coverage of time-varying graphs. In: Kratsch, D., Todinca, I. (eds.) WG 2014. LNCS, vol. 8747, pp. 29–41. Springer, Heidelberg (2014). http://www.univ-orleans.fr/lifo/evenements/WG2014/
3. Baumann, H., Crescenzi, P., Fraigniaud, P.: Parsimonious flooding in dynamic graphs. Distr. Comp. **24**(1), 31–44 (2011)
4. Bektas, T.: The multiple traveling salesman problem: an overview of formulations and solution procedures. OMEGA **34**(3), 209–219 (2006)
5. Bhadra, S., Ferreira, A.: Complexity of connected components in evolving graphs and the computation of multicast trees in dynamic networks. In: Pierre, S., Barbeau, M., An, H.-C. (eds.) ADHOC-NOW 2003. LNCS, vol. 2865, pp. 259–270. Springer, Heidelberg (2003)
6. Bui-Xuan, B., Ferreira, A., Jarry, A.: Computing shortest, fastest, and foremost journeys in dynamic networks. IJ Found. Comp. Sci. **14**(02), 267–285 (2003)
7. Casteigts, A., Flocchini, P., Mans, B., Santoro, N.: Deterministic computations in time-varying graphs: broadcasting under unstructured mobility. In: Calude, C.S., Sassone, V. (eds.) TCS 2010. IFIP AICT, vol. 323, pp. 111–124. Springer, Heidelberg (2010)
8. Casteigts, A., Flocchini, P., Quattrociocchi, W., Santoro, N.: Time-varying graphs and dynamic networks. IJPED **27**(5), 387–408 (2012)
9. Avin, C., Koucký, M., Lotker, Z.: How to explore a fast-changing world (Cover Time of a Simple Random Walk on Evolving Graphs). In: Aceto, L., Damgård, I., Goldberg, L.A., Halldórsson, M.M., Ingólfsdóttir, A., Walukiewicz, I. (eds.) ICALP 2008, Part I. LNCS, vol. 5125, pp. 121–132. Springer, Heidelberg (2008)
10. Choset, H.: Coverage for robotics - a survey of recent results. Ann. Math. Artif. Intell. **31**, 113–126 (2001)
11. Correll, N., Rutishauser, S., Martinoli, A.: Comparing coordination schemes for miniature robotic swarms: a case study in boundary coverage of regular structures. In: Khatib, O., Kumar, V., Rus, D. (eds.) Experimental Robotics. STAR, vol. 39, pp. 471–480. Springer, Heidelberg (2008)
12. Dynia, M., Korzeniowski, M., Schindelhauer, C.: Power-aware collective tree exploration. In: Grass, W., Sick, B., Waldschmidt, K. (eds.) ARCS 2006. LNCS, vol. 3894, pp. 341–351. Springer, Heidelberg (2006)
13. Dynia, M., Kutyłowski, J., der Heide, F.M., Schindelhauer, C.: Smart robot teams exploring sparse trees. In: Královič, R., Urzyczyn, P. (eds.) MFCS 2006. LNCS, vol. 4162, pp. 327–338. Springer, Heidelberg (2006)
14. Easton, K., Burdick, J.: A coverage algorithm for multi-robot boundary inspection. In: Proceedings of ICRA, pp. 727–734 (2005)
15. Fakcharoenphol, J., Harrelson, C., Rao, S.: The k-traveling repairman problem. ACM Trans. Algorithms **3**(4) (2007). http://dl.acm.org/citation.cfm?doid=1290672.1290677
16. Flocchini, P., Mans, B., Santoro, N.: On the exploration of time-varying networks. Theor. Comput. Sci. **469**, 53–68 (2013)
17. Garey, M., Johnson, D.: Computers and Intractability: A Guide to the Theory of NP-Completeness. W. H. Freeman, New York (1979)

18. Godard, E., Mazauric D.: Computing the dynamic diameter of non-deterministic dynamic networks is hard. In: Gao, J., Efrat, A., Fekete, S.P., Zhang, Y. (eds.) ALGOSENSORS 2014. LNCS, vol. 8847, pp. 88–102. Springer, Heidelberg (2015)

19. Ilcinkas, D., Wade, A.M.: On the power of waiting when exploring public transportation systems. In: Fernàndez Anta, A., Lipari, G., Roy, M. (eds.) OPODIS 2011. LNCS, vol. 7109, pp. 451–464. Springer, Heidelberg (2011)

20. Ilcinkas, D., Wade, A.M.: Exploration of the T-Interval-Connected Dynamic Graphs: the case of the ring. In: Moscibroda, T., Rescigno, A.A. (eds.) SIROCCO 2013. LNCS, vol. 8179, pp. 13–23. Springer, Heidelberg (2013)

21. Kuhn, F., Lynch, N., Oshman, R.: Distributed computation in dynamic networks. In: STOC, pp. 513–522 (2010)

22. Kuhn, F., Oshman, R.: Dynamic networks: models and algorithms. ACM SIGACT News **42**(1), 82–96 (2011)

23. Kumar, S., Lai, T., Arora, A.: Barrier coverage with wireless sensors. In: ACM MobiCom, pp. 284–298 (2005)

24. Mans, B., Mathieson, L.: On the treewidth of dynamic graphs. In: Du, D.-Z., Zhang, G. (eds.) COCOON 2013. LNCS, vol. 7936, pp. 349–360. Springer, Heidelberg (2013)

25. Michail, O., Spirakis, P.G.: Traveling salesman problems in temporal graphs. In: Csuhaj-Varjú, E., Dietzfelbinger, M., Ésik, Z. (eds.) MFCS 2014, Part II. LNCS, vol. 8635, pp. 553–564. Springer, Heidelberg (2014)

26. Nagamochi, H., Okada, K.: A faster 2-approximation algorithm for the minmax p-traveling salesmen problem on a tree. Discrete Applied Math. **140**(1-3), 103–114 (2004)

27. Xu, L., Xu, Z., Xu, D.: Exact and approximation algorithms for the minmax k-traveling salesmen problem on a tree. EJOR **227**, 284–292 (2013)

Strategies for Parallel Unaware Cleaners

Christian Ortolf[✉] and Christian Schindelhauer

Department of Computer Science, Computer Networks, University of Freiburg,
Freiburg im Breisgau, Germany
{ortolf,schindel}@informatik.uni-freiburg.de

Abstract. We investigate the parallel traversal of a graph with multiple robots unaware of each other. All robots traverse the graph in parallel forever and the goal is to minimize the time needed until the last node is visited (*first visit time*) and the time between revisits of a node (*revisit time*). We also want to minimize the *visit time*, i.e. the maximum of the first visit time and the time between revisits of a node. We present randomized algorithms for uncoordinated robots, which can compete with the optimal coordinated traversal by a small factor, the so-called *competitive ratio*.

For ring and path graph simple traversal strategies allow constant competitive factors even in the worst case. For grid and torus graphs with n nodes there is a $\mathcal{O}(\log n)$-competitive algorithm for both visit problems succeeding with high probability, i.e. with probability $1 - n^{-\mathcal{O}(1)}$. For general graphs we present an $\mathcal{O}(\log^2 n)$-competitive algorithm for the first visit problem, while for the visit problem we show an $\mathcal{O}(\log^3 n)$-competitive algorithm both succeeding with high probability.

Keywords: Visit time · Competitive analysis · Mobile agent · Robot · Multi-robot graph exploration

1 Introduction

Today, we are used to robotic lawn mowers and robotic vacuum cleaning. The current best-selling technology relies on robots which have no communication features and in some cases use maps of the environment. If we model the environment as an undirected graph, then its traversal by a single robot is an NP-hard minimum Traveling Salesman problem, for which efficient constant factor approximation algorithms are known [3]. Now, the robot owner deploys additional robots. How well do these robots perform? Can we guarantee that two parallel unaware lawn mowers will cut all grass better than one? And how do they compare to a couple of perfectly choreographed mowers? What about more robots, where each robot has no clue how many co-working devices exist nor where they are?

Here, we investigate these questions. We model the cleaning area by a graph with identifiable nodes and edges. All robots know only their own position and the graph. They will never learn how many robots are involved, nor any other

© Springer-Verlag Berlin Heidelberg 2015
J. Gao et al. (Eds.): ALGOSENSORS 2014, LNCS 8847, pp. 39–56, 2015.
DOI: 10.1007/978-3-662-46018-4_3

robots' positioning data. So, we assume that robots pass each other on the same node without noticing. We are looking for a traversal strategy of the graph which is self-compatible, since we assume that all robots are clones performing the same strategy.

It appears apparent that such a strategy must be probabilistic, since robots starting from the same node would otherwise follow identical routes, which would not allow for any speedup. However, we will see that this is not the case for all graphs.

Related Work. To our knowledge this unaware parallel cleaning model is new, therefore we will point out similarities to other problems.

The parallel unaware cleaning can be seen as a variation of the multi robot exploration [4–6,8,15]. The goal of the online multi-robot exploration is to steer a group of robots to visit every node of an unknown graph. The term *unknown* means that the exploring algorithm knows only edges adjacent to formerly visited nodes. The performance of such online algorithms is usually provided by a competitive analysis comparing the online solution to the optimal offline strategy, where an algorithm is given knowledge of the whole graph beforehand. This model is close to our first visit time model with two important differences: In parallel unaware cleaning each robot knows the full graph, while multi-robot exploration robots know only the explored graph. In our model there is no communication, while in robot exploration robots exchange their graph information.

It was recently shown that if more than dn robots are used in the multi-robot exploration problem, where d is the diameter of the graph and n the number of nodes, then one can achieve a constant competitive factor for multi-robot exploration [4]. The competing offline exploration can explore a graph in time $\Theta(\frac{n}{k} + d)$, therefore an exploration using $k = \frac{n}{d}$ robots is of special interest, because it allows the offline algorithm to make full use of all its robots.

For this scenario Dynia et al. [6] showed the online exploration of trees to be at best $\Omega(\frac{\log k}{\log \log k})$-competitive. If algorithms are restricted to greedy exploration an even stronger bound of $\Omega(k/\log k)$ is shown by Higashikawa et al. [11]. This bound matches the best known upper bound by Fraigniauds et al.'s greedy exploration algorithm in [8]. For further restricted graphs better bounds have been shown. An algorithm depending on a *density* parameter p was presented by Dynia et al. [5] with $\mathcal{O}(d^{1-1/p})$ competitiveness, e.g. $\mathcal{O}(d^{1/2})$ for trees embeddable in grids. For grids with convex obstacles, an polylogarithmic competitive bound of $\mathcal{O}(\log^2 n)$ was shown in [15], along with the lower bound of $\Omega(\frac{\log k}{\log \log k})$ matching the identical lower bound for trees.

Our problem also bears resemblance to the multi traveling salesman problem (mTSP) [2,9], a generalization of the well-known traveling salesman problem (TSP) [12]. TSP is NP-hard even in the seemingly simpler Euclidean version [16], but can be efficiently approximated if it is allowed to visit nodes more than once [18].

The mTSP tries to cover the graph with a set of tours and minimize the length of the longest tour. This corresponds to the offline parallel cleaning problem, if we use the distance between nodes in the graph as cost measure between nodes in mTSP. Even if salesmen start at different nodes the problem can still be reduced to the regular mTSP [10].

A similar definition to our first visit time is the notion of *cover time* for random walks, likewise visit time can be compared to the *hitting time* $H(i,j)$, the expected time starting from node i to reach node j. Our robots are not forced to use random walks. So, the Lollipop graph, a lower bound construction for the cover time of $\Omega(n^3)$ [13] and obtained by joining a complete graph to a path graph with a bridge, can be cleaned quite efficiently by parallel unaware cleaners.

Patrolling algorithms [17] also require robots to repeatedly visit the same area. To the best of our knowledge no work there has similarly restricted robots.

2   Model

In our model k robots are initially positioned on depot/starting nodes $S = (s_1, ..., s_k)$ and their task is to visit all nodes V of an undirected connected graph $G = (V, E)$ and then repeat their visits as fast as possible. Nodes and edges can be identified and time is measured in rounds. An algorithm has to decide for each robot r in each round which edge to traverse to visit another node in the following round. This decision is based on the starting node s_r, the graph and the previous decisions of the robot. Each robot never learns the number and positions of other robots.

The **first visit time of a node** is the number of the round, when a robot visits this node for the first time. The **visit time of a node** is the supremum of all time intervals between any two visits of a node (revisit) including the time interval necessary for the first visit by any robot. The **long term visit time of a node** is the supremum of time intervals between any two visits of a node by any robot after an arbitrarily long time. The corresponding definitions for the full graph is given by the maximum (first/long term) visit time of all nodes. Note that the robots do neither know and nor are they able to compute the visit times. These times can only be known by an external observer.

The term **with high probability** refers to an event which occurs with probability $1 - n^{-c}$ with constant $c \geq 1$. In all of our results, this constant c can be arbitrarily increased if one allows a larger constant factor for the run-time.

The term **distance** refers to the number of edges on a shortest path between two nodes.

The benchmark for our solution is the time of an algorithm with full knowledge, i.e. the number and positions of all robots. The quotient between the unaware visit time and the full knowledge visit time is our measure, also known as the competitive factor. The worst case setting can be seen as an adversary placing the robots for a given algorithm.

3   Simple Cleaning Examples

As an illustration and starting example we show how differently a circle graph and a path graph behave in this setting, see Figs. 1 and 2. The simple algorithm

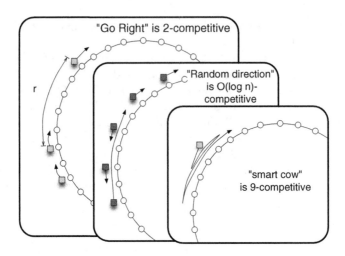

Fig. 1. Parallel unaware cleaning algorithms for the cycle graph. Illustrating competitive ratio for first visit.

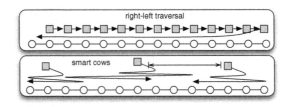

Fig. 2. Parallel unaware cleaning algorithms for the Path graph

sending robots in one direction in the circle, or just to one end on the line, then returning to the other end, performs quite differently for both graphs.

On the circle the right traversal strategy performs very well, the first visit time may be improved by a knowing algorithm at most by a factor of 2, since the largest distance r between two robots at the beginning lowerbounds the optimal offline strategy by $r/2$. The deterministic right traversal strategy on the cycle visits all nodes in r rounds for the first round and revisits them in this frequency thereafter.

For the path graph, the overhead of such an algorithm is a factor of n. If one end node is not covered and all robots walk first to the right end and then return, no robot can visit the left node in less than n rounds. A smarter oblivious algorithm could improve this by sending robots into a random direction instead, yielding a competitive factor of $\mathcal{O}(\log n)$ in the expectation. However, a deterministic solution exists: the smart cow algorithm [1], which in the i-th phase for $i = 1, 2, \ldots, n$ explores 2^i nodes first to the left and then 2^i nodes to right from the starting node. While the smart cow algorithm is designed to find a hole in a fence, which it does within a competitive factor of nine, the same competitive factor can be shown for the cycle and the path graph. This shows that for these simple graphs deterministic competitive visiting strategies exist.

However, for the long term visit problem the situation is different. Symmetry cannot be resolved by any deterministic algorithm. If all robots have the same starting node no competitive ratio better than $\mathcal{O}(n)$ can be achieved for these algorithms. The following chapter shows a simple solution to the long term visit problem.

4   Canonical Cleaning and General Observations

Now we present first general strategies and techniques. For $u \in V$ let $N_\ell(u)$ denote the set of nodes in G within distance of at most ℓ to the node u. For a set $A \subseteq V$ let $N_\ell(A) = \bigcup_{u \in A} N_\ell(u)$. The following lemma is the key technique, which provides a lower bound for the number of robots in the vicinity.

Lemma 1. *Given a graph with a robot placement with a first visit time of t_f. Then, for any set of nodes A the number of robots in the node set $N_\ell(A)$ is at least $\lceil |A|/(t_f + 1) \rceil$ for $\ell \geq t_f$.*

Proof. First note that for each cleaning strategy it is not possible that robots outside of $N_{t_f}(A) \subseteq N_\ell(A)$ can reach any node within A in at most t_f steps. Let k be the number of robots that explore A within time t_f. At the beginning at most k nodes can be occupied by k robots. Then, in every subsequent round at most k additional nodes of A can be visited. In order to visit all nodes in A we have $k(t_f + 1) \geq |A|$. This implies $k \geq \frac{|A|}{t_f + 1}$. $\qquad\blacksquare$

Later on, we use this lemma in a bait-and-switch strategy. We use A as bait to ensure that enough robots exist in a region for the offline strategy. Then we switch and let these robots work on other areas.

While randomization is necessary for dispersing the robots, too many probabilistic decisions are problematic, because the chance that some nodes remain unvisited for long times may grow over time. Therefore, we present only algorithms that use a finite number of randomized decisions. This technique is presented in the canonical algorithm, which is the base for some of our strategies. It requires the algorithms *cycle-start-node* and *waiting time* to provide where and when the robot should start cycling the graph (Fig. 3).

Because of the coupon collector's problem, a basic problem of probability theory [14], one cannot expect a better competitive factor than $\mathcal{O}(\log n)$. Therefore, in the long run the problem can be solved by the canonical algorithm.

Theorem 1. *Using the* CANONICAL CLEANING *it is possible to achieve a long-term visit time of $\mathcal{O}((n/k) \log n)$ and a visit time of $\mathrm{diameter}(G) + \mathcal{O}((n/k) \log n)$ with high probability.*

We refer to the Appendix A.1 for the proof.

For graphs with small diameter this results in a logarithmic competitive ratio. E.g. in balanced trees the diameter is bounded by $\mathcal{O}(\log n)$. So, the CANONICAL CLEANING algorithm gives us the following bound.

Algorithm 1. CANONICAL CLEANING algorithm for robot r using algorithms cycle-start-node and waiting-time

---

Traverse the graph by DFS yielding a cycle P with $V(P) = V$ of length $2n$
$v_s \leftarrow$ cycle-start-node(s_r)
Move robot r on the shortest path to v_s
$w \leftarrow$ waiting time(s_r, v_s)
Wait w rounds
if v_s *occurs more than once in* P **then**
| Choose a random occurrence in P
end
while true do
| Walk to the next node of P
end

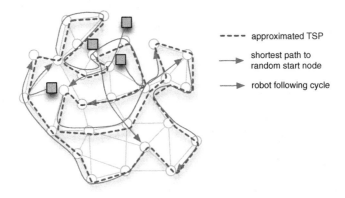

Fig. 3. A canonical algorithm guarantees a $\mathcal{O}(\frac{n}{k} \log n)$ long-term visit time.

Corollary 1. *Graphs with diameter of $\mathcal{O}(\log n)$ have a competitive ratio of $\mathcal{O}(\log n)$ for the first and revisit visit time with high probability.*

Proof. Let *cycle-start-node*(u) map to a uniform random node v of the tree. And let *waiting-time*(u, v) = *diameter*(G) $- |u, v|$. Let t_f^* and t_v^* be the optimal first and visit times and let $k \leq n$ be the number of robots.

Theorem 1 states that the first visit and visit time is bounded by *diameter* $(G) + \mathcal{O}((n/k) \log n) = \mathcal{O}(\log n + (n/k) \log n) = \mathcal{O}((n/k) \log n)$. From Lemma 1 it follows for $A = V$ that $t_f^* \geq n/k - 1$ and $t_v^* \geq n/k$. This implies a competitive ratio of $\mathcal{O}(\log n)$ for $k \leq n$. If $t_f^* > 0$ it also holds for $k \geq n$. In the case of $t_f^* = 0$, the robots already cover all nodes and every algorithm is optimal for the first visit time.

Another interesting technique is to transform a probabilistic first visit time strategy into a visit time algorithm succeeding with high probability. The only drawback is, that the first visit time and the visit probability for all nodes must be known.

Lemma 2. *Assume there exists a parallel unaware cleaner algorithm $\mathcal{A}$ for k robots on a graph with n nodes, where for all nodes u the probability that the first visit time is less or equal than t_f is at least $p > 0$. Furthermore, t_f and p are **known**. Then, this cleaning algorithm can be transformed into a canonical algorithm having visit time $\mathcal{O}(\frac{1}{p} t_f \log n)$ with high probability.*

The proof sketch is the following. Let $P(r)$ with $|P(r)| \leq t_f$ be the resulting path of robot r performing algorithm $\mathcal{A}$. Then, the *cycle-start-node* of the canonical algorithm is defined by choosing a random uniform node v_s from $P(r)$. We use *waiting-time* $(s_r, v_s) = 0$. In the Appendix A.2 a detailed proof is given.

5   The Torus and the Grid Graph

Now we consider torus and grid graphs, where we present optimal unaware cleaner strategies.

Define a $m \times m$-Torus $G_T = (V, E_T)$ graph by $V = [0, \ldots, m-1] \times [0, \ldots, m-1]$ and with edges $\{(i,j), (i+1 \mod m, j)\}$ and $\{(i,j), (i, j+1 \mod m)\}$ for $(i,j) \in V$. Every node has four neighbors, where we call the directions *right*, *left*, *up*, and *down* in the standard way. Parallel unaware robots can clean the torus graph with only a small overhead.

Algorithm 2. Competitive torus cleaner strategy for robot r

$(x, y) \leftarrow (s_{r.x}, s_{r.y})$ starting position
for $i \leftarrow 1, 2, \ldots, \sqrt{n}$ **do**
$\quad$ **if** *random event occurs with probability* $(x - s_{r.x} + 1)/(i+1)$ **then**
$\quad\quad$ $x \leftarrow x + 1$
$\quad$ **else**
$\quad\quad$ $y \leftarrow y + 1$
$\quad$ **end**
$\quad$ Move to (x, y)
end
$H := $ cycle of Fig. 5.
while true do
$\quad$ Move to the next node of H
end

Theorem 2. *Algorithm 2 is a high probability $\mathcal{O}(\log n)$-competitive visit algorithm for the $m \times m$-torus graph.*

We refer to the Appendix A.3 for analysis of Algorithm 2.

The first technique, the for loop of Algorithm 2, is that the cleaner uses a probabilistic process to create a uniform probability distribution over a linear growing and moving set of diagonal nodes. A pure random walk would create a binomial distribution. So, the probability distribution "pushes" to the corners, see Fig. 4.

Likewise in the canonical algorithm we switch after some time to a deterministic cycling algorithm. The difference is, that this cycle is adapted to the first phase and is a perfect Hamiltonian cycle, see Fig. 5.

The proof relies on the bait-and-switch-strategy, where the bait is a diagonal field of length t and width $2t_f$. In the neighborhood of such a field at least $\Omega(t)$ robots must be placed at the beginning or the offline strategy does not succeed within first visit time t_f. The first phase of the cleaner strategy moves these robots to a given target node with probability $\mathcal{O}(1/t)$. So, a constant number of robots pass any target node within any time frame of length $\mathcal{O}(t_f)$. Since, the robots' random decisions are independent an increase of a factor of $\mathcal{O}(\log n)$ gives the time bound for the first phase.

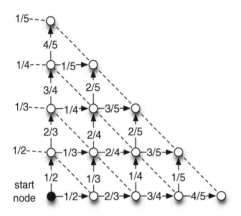

Fig. 4. Torus cleaner strategy   **Fig. 5.** Final cycle through the torus

For the second cycling phase, we have chosen the cycle with respect to the first phase, such that the same argument can be reused in order to estimate the maximum distance between two nodes on this cycle. The full proof can be found in the Appendix A.3.

This algorithm can be easily adapted for the grid graph, which consists of the same node set, but edges $\{(i,j),(i+1,j)\}$ for $i \neq m, (i,j) \in V$ and $\{(i,j),(i,j+1)\}$ for $j \neq m, (i,j) \in V$.

Theorem 3. *There exists a high probability $\mathcal{O}(\log n)$-competitive visit time cleaning algorithm for the $m \times m$-grid graph with $n = m^2$ nodes.*

Proof. We embed a $2m \times 2m$-torus graph G_T on the $m \times$-grid graph G_G by mapping the four nodes $(x,y), (2m-x+1,y), (x,2m-y+1), (2m-x+1,2m-y+1)$ onto the node $(x,y) \in V(G_G)$. Note that the edges of the torus map to edges in the grid.

At the beginning we choose for a robot a random representative in the torus graph and then we follow the algorithm for the torus graph. The proof is analogous to the one of the torus graph presented in the appendix except to a constant factor increase of the competitive factor.

6   Unaware Parallel Traversal of General Graphs

For general graphs we use a partition of the graph, which balances the work load of the robots. For the randomized partition we are inspired by the techniques of embedding tree metrics for graphs [7].

We partition the graph into disjoint recruitment areas $R_1, \ldots, R_n \subseteq V$. All robots in a recruitment area R_i have to visit the nodes in a working area W_i which is a proper subset of R_i. These sets are defined by a random process such that each node has a constant probability to be contained in a working area and we show that the number of robots in the recruitment area is large enough to ensure that this node is visited with constant probability. This constant probability will be increased later on by repeating the partitioning several times.

We give a formal description of the sets used in Algorithm 3. The recruitment partition uses center nodes $c_1, \ldots, c_n$ which are given by a random permutation π of all nodes $V = \{v_1, \ldots, v_n\}$, i.e. $c_i = v_{\pi(i)}$. The partition is based on the neighborhood set $N_\ell(u)$, which is the set of nodes v for which the distance to u is at most ℓ. So, we define for a radius ℓ and for all $i \in \{1, \ldots, n\}$.

$$R_i := N_\ell(v_{\pi(i)}) \setminus \bigcup_{j=1}^{i-1} N_\ell(v_{\pi(j)}) . \tag{1}$$

The working areas are defined for radius ℓ and an estimation of the first visit time $t \in [t_f, 2t_f]$ as

$$U_i := N_{l-2t}\left(v_{\pi(i)}\right) \setminus \bigcup_{j=1}^{i-1} N_{\ell+2t}\left(v_{\pi(j)}\right) \tag{2}$$

$$W_i := N_t\left(U_i\right) \tag{3}$$

We denote by $W = \bigcup_{i=1}^{n} W_i$ the set of nodes that will be worked on and let $U := \bigcup_{i=1}^{n} U_i$.

These definitions are used for a probabilistic cleaning Algorithm 3, which covers a constant part of the graph. The ONE-SHOT-CLEANING algorithm makes use of an straight-forward constant factor Steiner-tree approximation based on Prim's minimum spanning tree algorithm, presented as Algorithm 4.

The following lemma shows that every node is chosen with probability of at least $\frac{1}{4}$ to be the target of a robot cleaning in some area W_i.

Lemma 3. *For a graph G, a node $v \in V$, β chosen randomly from $[1, 2]$, a random permutation π over $\{1, \ldots, n\}$, and for $l = 8\beta t \log n$ the probability that $v \in W$ is at least $\frac{1}{4}$.*

We refer to the Appendix A.4 for the proof.

Algorithm 3. ONE-SHOT-CLEANING $G = (V, E)$ using $V = R_1 \dot{\cup} \cdots \dot{\cup} R_n$
and $W_1, \ldots, W_n \subseteq V$

Choose i such that $s_r \in R_i$
$T_i \leftarrow$ STEINER-TREE-APPROXIMATION(W_i)
$C_i \leftarrow$ DFS-Cycle(T_i)
Move to a random node of C_i
Walk on C_i for $68t \log n$ rounds
Move back to s_r

Algorithm 4. STEINER-TREE-APPROXIMATION with input $G = (V, E)$,
$W \subseteq V$

$(C_1, \ldots, C_p) \leftarrow$ connected components of W in G
while $p > 1$ **do**
    Choose the component C_j with the nearest node to C_1
    $W \leftarrow W \cup$ (node set of shortest path between C_1 and C_j to W)
    $(C_1, \ldots, C_p) \leftarrow$ connected components of W
end
return spanning tree of C_1

Now, we investigate whether there are enough robots in the recruitment area R_i in order to explore W_i. The number is large enough if a given node is explored with a constant probability. However, there is a major problem: U_i, W_i, or R_i might be disconnected. So robots might travel long routes between the nodes in W_i outside of W_i or even R_i.

Therefore, we need an upper bound on the size of these connecting routes. This has been the motivation to extend U with a surrounding of t neighborhood nodes. So, for $\beta \in [1, 2]$ we have the following lemma.

Lemma 4. *For $\ell = 8\beta t \log n$, let T_i be the tree connecting all nodes in W_i constructed in Algorithm 4. Then,*

$$|V(T_i)| \leq 17|W_i| \log n .$$

Proof. Each of the p connected components $C_1, \ldots, C_p$ of W_i has at least one node of U and its t-neighborhood. So, C_j has at least t nodes, implying $|W_i| \geq pt$. Every node of W_i has distance of at most $\ell = 8\beta t \log n$ to $v_{\pi(i)}$. The maximum distance between two components is thus at most $16\beta t \log n$ because of the triangle inequality. Which implies that at most $16(p-1)\beta t \log n$ nodes are added to connect the original p connected components. So,

$$|V(T_i)| \leq 16(p-1)\beta t \log n + |W_i|$$
$$\leq 16 \frac{p-1}{p} |W_i| \log n + |W_i|$$
$$\leq 17|W_i| \log n .$$

The following lemma shows that the ONE-SHOT-CLEANING algorithm needs only a logarithmic overhead.

Lemma 5. *The number of moves of a robot using* ONE-SHOT-CLEANING *for* $\ell = 8\beta t \log n$ *and* $\beta \in [1, 2]$ *is at most* $100t \log n$.

Proof. The maximum distance of any node from u to W_i is at most $\ell - t = 8\beta t \log n - t \leq 16t \log n$. So, moving to the start node and moving back to the start node needs at most $32t \log n$ rounds. Moving on C_i needs $68t \log n$ rounds resulting in $100t \log n$ rounds.

Now, we need to show that the number of robots in the recruitment area R_i is large enough. This follows by Lemma 1 substituting $A = W_i$.

Lemma 6. *If the robots are placed such that a first visit time of* t_f *is possible, and* $t \in [t_f, 2t_f]$, *then for the number* k_i *of robots originally placed in* R_i *we have*

$$k_i \geq \frac{|W_i|}{t_f + 1} \geq \frac{|W_i|}{2t}.$$

Proof. A single robot can explore at most $t_f + 1$ nodes in the first t_f rounds. Therefore the minimum amount of nodes to be explored by all robots in R_i is $k_i(t_f + 1) \leq 2k_i t_f$.

These observations allows us to find a general strategy for the first visit problem for unaware parallel cleaners.

Algorithm 5. High probability first visit cleaner of $G = (V, E)$

---

for $i \in \{1, 2, \ldots, \log n\}$ **do**
  $t \leftarrow 2^i$
  for $j \in \{1, \ldots, 4(c + 1) \ln n\}$ **do**
    Choose randomly $\beta \in [1, 2]$
    Choose random permutation π over V
    ONE-SHOT-CLEANING$(G, \ell = 8\beta t \log n, t, \pi)$
  end
end

Theorem 4. *Algorithm 5 is a high probability* $\mathcal{O}(\log^2 n)$*-competitive first visit algorithm for every undirected graph.*

Repeating the ONE-SHOT-CLEANING $\mathcal{O}(\log n)$ times gives us high probability. The full proof can be found in Appendix A.5.

The visit time problem needs more moves, since a robot may make a fast first visit, but does not know when to end. Our solution is to guess the first visit time.

Algorithm 6. High probability visit of $G = (V, E)$

Choose uniform at random $i \in \{1, 2, \ldots, \log n\}$
$t \leftarrow 2^i$
Choose randomly $\beta \in [1, 2]$
Choose random permutation π over V
ONE-SHOT-CLEANING(G, t, β, π)
Traverse the graph by DFS yielding a cycle C with $V(C) = V$ of length $2n$
Go to a random node visited during the one shot cleaning
while true do
| Walk to the next node of C
end

Theorem 5. *Algorithm 6 is an high probability $\mathcal{O}(\log^3 n)$-competitive visit algorithm for every undirected graph.*

Proof. Lemma 3 implies that $P(w \in W_i) \geq \frac{1}{4}$ if $\ell = 8\beta t \log n$. The probability that a robot chooses the correct value $t = 2^i \in [t_f, 2t_f]$ is $1/\log n$. So, the probability that a node is visited within first visit time $800ct_f \log n$ is at least $p = \frac{1}{4 \log n}$. By Lemma 2 this implies a visit time algorithm with high probability with time $\mathcal{O}(t_f \log^3 n)$. □

7   Conclusion

We discuss a central question of distributed algorithms: How much do we benefit from communication? Or to put it otherwise: Can we cope with a parallel problem if communication is not available? We have shown that first visit can be achieved with an overhead of $\mathcal{O}(\log^2 n)$ and visit with $\mathcal{O}(\log^3 n)$ in general graphs. This means that we can cope quite well without any communication.

For the grid and torus we show an even stronger bound of $\mathcal{O}(\log n)$. This matches the lower bound of $\Omega(\log n)$ given by the coupon collector's problem. Unlike the algorithm presented for general graphs the parallel unaware cleaner strategy for torus and grids have only small constant factors involved. Furthermore, the grid represents a typical application areas for such robots. So, we can very well envisage our cleaning strategies to be implemented onto current room cleaning and lawn mowing robots.

A   Appendix

A.1   Canonical Cleaning

Theorem 1. *Using the CANONICAL CLEANING it is possible to achieve a long-term visit time of $\mathcal{O}((n/k) \log n)$ and a visit time of diameter$(G) + \mathcal{O}((n/k) \log n)$ with high probability.*

Proof. We choose for each robot an independent uniform random choice of the nodes of the cycle P as the *cycle-start-node*. The *waiting-time* is defined as $diameter(G) - |s_r, v_s|$. So, all nodes start the traversal at the same time.

Let g be a subpath on the cycle P of length at most $2n$. The probability that no robots are in this subpath is $(1 - \frac{g}{|P|})^k$. For k robots a subpath $g \geq \frac{2cn \ln n}{k}$ is empty with probability

$$\left(1 - \frac{g}{|P|}\right)^k \leq \exp\left(-\frac{gk}{|P|}\right) \leq \exp\left(-\frac{gk}{2n}\right) \leq \exp\left(-c \ln n\right) \leq n^{-c} .$$

Hence, the maximum gap between two nodes on the cycle P is at most $\mathcal{O}((n/k) \log n)$ with high probability.

So, the long term visit time is bounded by this gap. From the waiting time, the first visit time follows. Note that after the first visit, the revisit time matches the long term visit time.

A.2   Canonical Algorithm First Visit

Lemma 2. *Assume there exists a parallel unaware cleaner algorithm $\mathcal{A}$ for k robots on a graph with n nodes, where for all nodes u the probability that the first visit time is less or equal than t_f is at least $p > 0$. Furthermore, t_f and p are **known**. Then, this cleaning algorithm can be transformed into a canonical algorithm having visit time $\mathcal{O}(\frac{1}{p}t_f \log n)$ with high probability.*

Proof. Let $P(r)$ with $|P(r)| \leq t_f$ be the resulting path of robot r performing algorithm $\mathcal{A}$. Then, the *cycle-start-node* of the canonical algorithm is defined by choosing a random uniform node v_s from $P(r)$. We set *waiting-time* $(r)=0$.

We now show that this algorithm fulfills the time behavior.

1. The first visit time can be proved as follow.

   Each node is visited with probability of at least $\frac{p}{t_f}$. However, there are dependencies between these events, since nodes might be visited by the same robot. So, we consider the subpath before a node v of length $\frac{2ct_f \ln n}{p}$ on a cycle C of length $2n$ with $V(C) = V$. Then, at least $c \ln n$ different robots have positive probabilities to visit this interval. Let $1, \ldots, k$ be these robots and let p_i be the probability that one of these robots visits this interval. For these probabilities we have $\sum_{i=1}^{k} p_i \geq \frac{p}{t_f} \frac{ct_f \ln n}{p} = c \ln n$, since otherwise a node exists which is visited with smaller probability than $\frac{p}{t_f}$.

   The probability for not visiting this interval is therefore

$$\prod_{i=1}^{k} (1 - p_i) \leq \prod_{i=1}^{k} \exp\left(-p_i\right) \leq \exp\left(-\sum_{i=1}^{k} p_i\right) \leq \exp\left(-c \ln n\right) \leq n^{-c} .$$

   Since with high probability a cycle-start-node is chosen on the cycle P at most $(2ct_f \ln n)/p$ nodes before v, v will be visited after $t_f + 2\frac{c}{p}t_f \ln n$ steps for the first time w.h.p. From the union bound the claim follows.

2. The visit time follows by the following observation: From the observations above we know that the subpath of length $2ct_f \ln n$ on P before and after any node is visited within time t_f. Therefore the visit time of a node is at most $4ct_f \ln n + 2t_f$.

A.3   Analysis of Torus Algorithm

Theorem 2. *Algorithm 2 is a high probability $\mathcal{O}(\log n)$-competitive visit cleaning algorithm for the $m \times m$-torus graph.*

Proof. The following Lemma shows that the torus algorithm distributes the robots with equal probabilities.

Lemma 3. *For all $t \in \{1, \ldots, \sqrt{n}\}$, $i \in \{0, \ldots, t\}$ the probability that a robot starting at node $(s_{r.x}, s_{r.y})$ is at node $(s_{r.x} + i, s_{r.y} + (t - i))$ after t rounds is $1/(t + 1)$.*

Proof. This follows by induction. For $t = 0$ the probability is 1 that the robot is at the start node $(s_{r.x}, s_{r.y})$. Assume that at round $t - 1$ the claim is true.

For the induction we have to consider three cases:

- If $x = s_{r.x}$ and $y = s_{r.y} + t$ then the probability to move to this point is the product of the stay probability at $(x, y - 1)$ and the probability to increment y. By induction this is $\frac{1}{t}\left(1 - \frac{1}{t+1}\right) = \frac{1}{t+1}$.
- If $y = s_{r.y}$ and $x = s_{r.x} + t$ then the probability to move to this point is the product of the stay probability at $(x, y - 1)$ and the probability to increment x. By induction this is again $\frac{1}{t}\left(1 - \frac{1}{t+1}\right) = \frac{1}{t+1}$.
- For all other cases we have to combine the probability to increment x and y, the sum of which is $\frac{t}{t+1}$. By induction we get as probability $\frac{1}{t}\frac{t}{t+1} = \frac{1}{t+1}$ claim follows.

Assume that t_f is the first visit time time for a robot placement in the torus. For the cleaning of a target node (x, y) we choose a set of nodes S with $t - 4t_f$ nodes at a diagonal in distance t, see Fig. 6. $A = N_{t_f}(S)$ is now the bait, i.e. the area, which guarantees the minimum number of robots the recruitment area $N_{t_f}(A)$. Lemma 1 states that at least $|A|/(t_f + 1)$ robots must be in this recruitment area $N_{t_f}(A)$. Now, the cleaning algorithm makes sure that all these robots pass through the target node during the time interval $[t - 2t_f, t + 2t_f]$ with a probability of at least $1/(t + 2t_f + 1)$. Now, the size of $|A|$ is at least $2t_f(t - 4t_f)$. So, the expected number of robots passing through the target node is at least

$$\frac{|A|}{(t_f + 1)(t + 2t_f + 1)} \geq \frac{2t_f(t - 4t_f)(t + 2t_f + 1)}{t_f + 1} \geq \frac{t - 4t_f}{t + 2t_f + 1}.$$

So for $t \geq 10t_f$ we expect at least a constant number of $\frac{1}{2}$ robots passing through any node in a time interval of length $3t_f$. If we increase the time interval

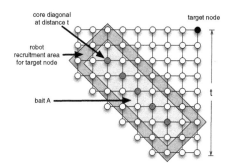

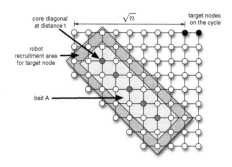

Fig. 6. The robot recruitment area for robots exploring the target node.

Fig. 7. The robot recruitment area for robots on the cycle.

to the size of some $ct_f \log n$ for some appropriately chosen constant c, applying a Chernoff bound ensures us to visit this node with at least one robot with high probability.

This proves that in the first phase of the algorithm we visit (and revisit) each node in every time intervals of length $\mathcal{O}(t_f \log n)$.

It remains to show that in the second phase, where the algorithm enters the cycle the distance on the cycle is bounded by $\mathcal{O}(t_f \log n)$. For this, we consider $4t_f < \sqrt{n}$ consecutive nodes on the cycle, which lie on $4t_f$ consecutive diagonals, see Fig. 7. So, all of the $|A|/(t_f + 1)$ robots in the recruitment area have a target node, which can be reached after $\sqrt{n}$ steps. For each of these target nodes, the probability to be reached by a robot on the corresponding diagonal is at least $\frac{1}{\sqrt{n}}$. The minimum size of $|A|$ is at least $\sqrt{n} - 2t_v$, which results in an expected number of at least

$$\frac{2t_f(\sqrt{n} - 2t_f)}{(2t_f + 1)\sqrt{n}} \geq 1 - \frac{t_f}{\sqrt{n}}$$

robots on the target nodes of the cycle. For $t_f \leq \frac{1}{2}\sqrt{n}$ this means that the expected number of robots in an interval of length $4t_f$ is at least $\frac{1}{2}$. So, the longest empty interval has length of at most $\mathcal{O}(t_f \log n)$ by applying Chernoff bounds on $\mathcal{O}(\log n)$ neighbored intervals.

For $t_f \geq \frac{1}{2}\sqrt{n}$ we consider $\sqrt{n}$ consecutive nodes on consecutive diagonals. Every robot ends the first phase and starts the cycle within this interval with probability $\frac{1}{\sqrt{n}}$. The minimum number of robots to explore all n nodes is at least $\frac{n}{t_f+1}$, which follows by Lemma 1 for $A = V$. Now, for $c\frac{t_f}{\sqrt{n}} \log n$ neighbored intervals on the cycle each of length $\sqrt{n}$ the probability that a single robot chooses a node in this interval is at least

$$\frac{t_f}{\sqrt{n}} \frac{c \log n}{\sqrt{n}} = c\frac{t_f}{n} \log n \ .$$

So, the expected number of robots is $c\frac{n}{t_f}\frac{t_f}{n}\log n = c\log n$ for an time interval of length $c\frac{t_f}{\sqrt{n}}\sqrt{n}\log n = ct_f \log n$. Now, by Chernoff bounds the probability that we find this interval to be empty has a probability of at most $n^{-c'}$ for some constants c, c'.

So, the maximum distance of two robots on a cycle in the first and second phase is at most $\mathcal{O}(t_f \log n)$ with high probability. Since the visit time is at least the first visit time the competitive ratio of $\mathcal{O}(\log n)$ follows.

A.4   Proof of Lemma 3

Lemma 3. *For a graph G, a node $v \in V$, β chosen randomly from $[1,2]$, a random permutation π over $\{1,\ldots,n\}$, and for $\ell = 8\beta t \log n$ the probability that $v \in W$ is at least $\frac{1}{4}$.*

Proof. We will prove that $P(v \in U) \geq \frac{1}{4}$, which implies the claim because $U \subset W$.

Consider the first node w in the $\ell+2t$-neighborhood of v according to the random permutation π, i.e. $w = u_{\pi(i*)}$ where $i^* = \min\{i \mid |v, u_{\pi(i)}| \leq \ell+2t\}$. If w is closer than $\ell-2t$ to v, i.e. $|v, w| \leq \ell-2t$, then v is in the working area of w (and U), since no node with smaller index can be closer than w, i.e. $w \in U_{i*} \subseteq U$. On the other hand if this node is in the critical distance $|v, w| \in (\ell-2t, \ell+2t]$, then it is excluded from U_{i*} and since i^* has the smallest index in the vicinity it is also not in any other working area, i.e. $v \notin U$. Since π is a random permutation the probability of $v \in W$ is given by the number of elements in the closer vicinity:

$$P_\ell(v \in U) = \frac{|N_{\ell-2t}(v)|}{|N_{\ell+2t}(v)|}$$

This implies

$$\prod_{i=0}^{2\log n} P_{\ell+4it}(v \in U) = \frac{|N_{\ell-t}(v)|}{|N_{\ell+8t\log n+2t}(v)|} \geq \frac{1}{n} \qquad (4)$$

Now, we choose β randomly from $\{1, 1+\frac{1}{2\log n}, 1+\frac{2}{2\log n}, \ldots, 1+\frac{2\log n-1}{2\log n}\}$ and compute $\ell = 8\beta t \log n$. Hence,

$$P(v \in U) = \frac{1}{2\log n} \sum_{i=0}^{2\log n-1} P_{8t\log n+4it}(v \in W)$$

Assume that $P(v \in U) < \frac{1}{4}$, then at least half of all values of $(P_{8t\log n+4it}(v \in W))_{i\in\{0,\ldots,2\log n-1\}}$ are smaller than $\frac{1}{2}$. Then, we observe the following.

$$\prod_{i=0}^{2\log n} P_{8t\log n+4it}(v \in U) < \left(\frac{1}{2}\right)^{\log n} = \frac{1}{n},$$

which contradicts (4). Therefore $P(v \in W) \geq P(v \in U) \geq \frac{1}{4}$.

The same argument holds, if we choose β randomly from the real interval $[1,2]$.

A.5   Analysis of Algorithm 5

Theorem 3. *Algorithm 5 is a high probability $\mathcal{O}(\log^2 n)$-competitive first visit algorithm for every undirected graph.*

Proof. Consider the round of the outer loop, where $t = 2^i \in [t_f, 2t_f]$, where t_f is the first visit time of the optimal algorithm. We show that in this round all nodes will be explored with high probability. Lemma 5 states that the number of robot moves of ONE-SHOT-CLEANING is bounded by $100 \cdot 2^i \log n$. So, the overall number of each robot moves is bounded by $800(c+1)\log^2 n$.

For any node u the probability, that the ONE-SHOT-CLEANING algorithm for $\ell = 8\beta t \log n$ chooses $u \in W$ is at least $\frac{1}{4}$ following Lemma 3. If u resides in W_i, the number of robots performing the cleaning is at least $|W_i|/(2t)$ implied by Lemma 6. These k_i robots have to explore a cycle of length at most twice the size of the connected Steiner-tree computed in Algorithm 4. These are at most $34|W_i| \log n$ nodes. Now, Algorithm 3 starts with a random node and explores $68t \log n$ nodes. So, after one execution of the ONE-SHOT-CLEANING algorithm the probability of a node not to be explored is at most

$$1 - \frac{1}{4} \frac{68t \log n}{34|W_i| \log n} = 1 - \frac{t}{2|W_i|}$$

The cleaning is be independently repeated for $k_i \geq \frac{|W_i|}{2t}$ times.

$$\left(1 - \frac{t}{2|W_i|}\right)^{\frac{|W_i|}{2t}} \leq e^{-\frac{1}{4}}$$

Hence, the maximum probability of a node not to be explored after $4(c+1)\ln n$ repetitions is at most $\frac{1}{n^c}$.

References

1. Baezayates, R., Culberson, J., Rawlins, G.: Searching in the plane. Inf. Comput. **106**(2), 234–252 (1993)
2. Bektas, T.: The multiple traveling salesman problem: an overview of formulations and solution procedures. Omega **34**(3), 209–219 (2006)
3. Christofides, N.: Worst-case analysis of a new heuristic for the travelling salesman problem. Technical report, DTIC Document (1976)
4. Dereniowski, D., Disser, Y., Kosowski, A., Pająk, D., Uznański, P.: Fast collaborative graph exploration. In: Fomin, F.V., Freivalds, R., Kwiatkowska, M., Peleg, D. (eds.) ICALP 2013, Part II. LNCS, vol. 7966, pp. 520–532. Springer, Heidelberg (2013)
5. Dynia, M., Kutyłowski, J., der Heide, F.M., Schindelhauer, C.: Smart robot teams exploring sparse trees. In: Královič, R., Urzyczyn, P. (eds.) MFCS 2006. LNCS, vol. 4162, pp. 327–338. Springer, Heidelberg (2006)
6. Dynia, M., Łopuszański, J., Schindelhauer, C.: Why robots need maps. In: Prencipe, G., Zaks, S. (eds.) SIROCCO 2007. LNCS, vol. 4474, pp. 41–50. Springer, Heidelberg (2007)

7. Fakcharoenphol, J., Rao, S., Talwar, K.: A tight bound on approximating arbitrary metrics by tree metrics. In: Proceedings of the Thirty-fifth Annual ACM Symposium on Theory of Computing, pp. 448–455. ACM (2003)
8. Fraigniaud, P., Gąsieniec, L., Kowalski, D.R., Pelc, A.: Collective tree exploration. Network **48**, 166–177 (2006)
9. Frederickson, G., Hecht, M.S., Kim, C.E.: Approximation algorithms for some routing problems. In: 17th Annual Symposium on Foundations of Computer Science, pp. 216–227, Oct 1976
10. GuoXing, Y.: Transformation of multidepot multisalesmen problem to the standard travelling salesman problem. Eur. J. Oper. Res. **81**(3), 557–560 (1995)
11. Higashikawa, Y., Katoh, N., Langerman, S., Tanigawa, S.-I.: Online graph exploration algorithms for cycles and trees by multiple searchers. J. Comb. Optim. **28**(2), 480–495 (2014)
12. Karp, R.M.: Reducibility Among Combinatorial Problems. Springer, Heidelberg (1972)
13. Lovász, L.: Random walks on graphs: a survey. In: Miklós, D., Sós, V.T., Szőnyi, T. (eds.) Combinatorics, Paul Erdős is Eighty, vol. 2, pp. 353–398. János Bolyai Mathematical Society, Budapest (1996)
14. Newman, D.J.: The double dixie cup problem. Am. Math. Mon. **67**, 58–61 (1960)
15. Ortolf, C., Schindelhauer, C.: Online multi-robot exploration of grid graphs with rectangular obstacles. In: Proceedings of the Twenty-fourth Annual ACM Symposium on Parallelism in Algorithms and Architectures, SPAA '12, pp. 27–36. ACM, New York (2012)
16. Papadimitriou, C.H.: The Euclidean travelling salesman problem is NP-complete. Theoret. Comput. Sci. **4**(3), 237–244 (1977)
17. Portugal, D., Rocha, R.: A survey on multi-robot patrolling algorithms. In: Camarinha-Matos, L.M. (ed.) Technological Innovation for Sustainability. IFIP AICT, vol. 349, pp. 139–146. Springer, Heidelberg (2011)
18. Rosenkrantz, D., Stearns, R., Lewis, P.: Approximate algorithms for the traveling salesperson problem. In: IEEE Conference Record of 15th Annual Symposium on Switching and Automata Theory, 1974, pp. 33–42, Oct 1974

Minimum-Traveled-Distance Gathering of Oblivious Robots over Given Meeting Points

Serafino Cicerone[1], Gabriele Di Stefano[1], and Alfredo Navarra[2]([⊠])

[1] Dipartimento di Ingegneria e Scienze dell'Informazione e Matematica,
Università degli Studi dell'Aquila, L'Aquila, Italy
{serafino.cicerone,gabriele.distefano}@univaq.it

[2] Dipartimento di Matematica e Informatica,
Università degli Studi di Perugia, Perugia, Italy
alfredo.navarra@unipg.it

Abstract. The paper considers a new variant of the gathering problem of oblivious and asynchronous robots moving in the plane. Robots operate in standard Look-Compute-Move cycles. In one cycle, a robot perceives the current configuration in terms of robots distribution (Look), decides whether to move toward some direction (Compute), and in the positive case it moves (Move). Cycles are performed asynchronously for each robot. Robots are anonymous and execute the same distributed algorithm that must guarantee to move all robots to meet at some point among a predetermined set. During the Look phase robots perceive not only the relative positions of the other robots, but also the relative positions of a set of points referred to as *meeting points* where gathering must be finalized.

We are interested in designing a gathering algorithm that solves the problem by also minimizing the total distances covered by all robots. We characterize when this gathering problem can be optimally solved, and we provide a new distributed algorithm along with its correctness.

1 Introduction

The gathering task is a basic primitive in robot-based computing systems. It has been extensively studied in the literature under different assumptions. The problem asks to design a distributed algorithm that allows a team of robots to meet at some common place. Varying on the capabilities of the robots as well as on the environment where they move, very different and challenging aspects must be faced (see [5,8,11,13] for a survey).

Robot Model. In this paper, we are interested in robots placed in the plane. Initially, no robots occupy the same location. Robots are equipped with sensors

---

Work partially supported by the Italian Ministry of Education, University, and Research (MIUR) under national research projects: PRIN 2012C4E3KT "AMANDA – Algorithmics for MAssive and Networked DAta" and PRIN 2010N5K7EB "ARS TechnoMedia – Algoritmica per le Reti Sociali Tecno- mediate".

J. Gao et al. (Eds.): ALGOSENSORS 2014, LNCS 8847, pp. 57–72, 2015.
DOI: 10.1007/978-3-662-46018-4_4

and motion actuators, and operate in *Look-Compute-Move* cycles (see, e.g. [8]).
The Look-Compute-Move model assumes that in each cycle a robot takes a
snapshot of the current global configuration (Look), then, based on the perceived
configuration, takes a decision to stay idle or to move toward a specific direction
(Compute), and in the latter case it moves (Move). The distance traveled in a
move is neither infinite nor infinitesimally small. More precisely, there exists an
(arbitrarily small) constant $\delta > 0$ such that if the destination point is closer
than δ, the robot will reach it, otherwise the robot will be closer to it of at least
δ. Cycles are performed asynchronously, i.e., the time between Look, Compute,
and Move operations is finite but unbounded, and it is decided by an adversary
for each robot. Moreover, during the Look phase, a robot does not perceive
whether other robots are moving or not. Hence, robots may move based on
outdated perceptions. In fact, due to asynchrony, by the time a robot takes a
snapshot of the configuration, this might have drastically changed. The scheduler
determining the Look-Compute-Move cycles timing is assumed to be fair, that
is, each robot performs its cycle within finite time and infinitely often.

Robots are assumed to be oblivious (without memory of the past), uniform
(running the same deterministic algorithm), autonomous (without a common
coordinate system, identities or chirality), asynchronous (without central coordination), without the capability to communicate.

During the Look phase, robots are assumed to perceive whether a same
location is occupied by more than one robot without acquiring the exact number.
This capability is usually referred to as *global weak multiplicity detection* [14].
From now we simply say robots are empowered with the *multiplicity detection*
capability. Another finite set of points from now on called *meeting points* is
detected by robots during the Look phase. Meeting points represent the only
locations where gathering can be finalized. As for robots, the meeting points are
detected as relative positions with respect to the robot performing a Look phase.

The aim of this work is to study the GATHERING OVER MEETING POINTS
problem, that is the design of a distributed gathering algorithm ensuring all
robots to reach the same location among those specified by the meeting points.
The rational behind this choice is twofold. From the one hand, we believe the
model is theoretical interesting, as it is a hybrid scenario in between the classical
environment where robots freely move in the plane (see, e.g., [2,3]), and the
more structured one where robots must move on the verticals of a graphs (see,
e.g., [6,10]), implemented here by the set of meeting points. A similar setting
but for the pattern formation problem has been considered in [9]. On the other
hand, meeting points for gathering purposes might be a practical choice when
robots move in particular environments where not all places can be candidate
to serve as gathering points.

This general problem is addressed here with an additional constraint: we
require the robots cover the *minimum total travel distance* to finalize the
gathering. This requirement introduces optimization aspects in the study of the
gathering problems on the plane, as already done for gathering on graphs in [7].

Our Results. In this paper, we fully characterize the new defined gathering problem in the plane. In particular, we first show that the constraint to gather robots at meeting points implies that some configurations result ungatherable. We characterize such configurations by introducing specific automorphisms of points on the plane.

Then, we observe that the constraint on the minimum overall distance involves the concept of *Weber point*, that is the point of the minimizing the distances from a given set of points in the plane. In the literature, it has been shown that if the provided points are not collinear, then the Weber point is unique [1, 4, 12], but its computation is in general unfeasible. In our context, it is possible that the Weber point does not belong to the set of meeting points, hence requiring to evaluate which meeting point minimizes the distances from robots. Moreover such a point can be not unique. We show that some configurations cannot be gathered on points that guarantee such a minimum, even though they are potentially gatherable.

We say that a gathering algorithm is *exact* if it ensures the robots to meet at some meeting point while covering the overall minimum distance. Hence, for all configurations admitting exact gathering, we provide a new distributed algorithm that always ensures robots to gather at one point among the meeting ones while minimizing the overall distance.

Finally, we remark that in the classic gathering problem on the plane robots can finalize the gathering everywhere and without constraints on the covered distances. In [3] it has been shown that the classic gathering problem on the plane is solvable for any $n > 2$, for any initial configuration (the problem is unsolvable for $n = 2$ [16]). However, in our setting, not all configurations containing only two robots are ungatherable.

Outline. The next section introduces the required notation and gives some basic definitions. Section 3 provides basic impossibility results. In particular, ungatherable configurations and configurations where exact gathering cannot be assured are identified. Section 4 provides our new gathering algorithm. It is presented in terms of few different strategies according to different types of configurations. The section also provides the correctness proof of the proposed algorithm. Finally, Sect. 5 concludes the paper. Due to space constraints, some technical details are omitted (they will be given in the full version of this paper).

2   Definitions

In this section we provide the basic concepts used throughout the paper.

Notation. The system is composed of a set $\mathcal{R} = \{r_1, r_2, \ldots, r_n\}$ of n mobile *robots*. At any time, the multiset $R_{\mathcal{R}} = \{p_1, p_2, \ldots, p_n\}$, with $p_i \in \mathbb{R}^2$, contains the *positions* of the robots in $\mathcal{R}$ (when no ambiguity arises, we shall omit the subscript of R). The set $U(R) = \{x \mid x \in R\}$ contains the *unique* robots' positions. F is a finite set of m distinct *meeting points* in the plane representing the only locations in which robots can be gathered. The pair $C = (R, F)$ represents a system *configuration*. A configuration C is *initial* at time t if at that time all robots have distinct positions (i.e., $U(R) = n$). A configuration C is *final* at time t if (i) at that time each robot computes or performs a null movement and (ii) there exists a point $f \in F$ such that $p_i = f$ for each $p_i \in R$; in this case we say that the robots have gathered on point f at time t. During the *Look* phase, we assume the *global weak multiplicity detection* property, i.e., robots can distinguish points that host more than one robot without acquiring the exact number. Additionally, in the same phase robots must be able to distinguish points hosting robots from points in F.

We study the GATHERING OVER MEETING POINTS problem (shortly, GMP), that is, the problem of transforming an initial configuration into a final one. The classical GATHERING problem does not take into consideration the set F, and hence any point $f \in \mathbb{R}^2$ can be used to gather the robots. A gathering algorithm is a deterministic algorithm that brings the robots in the system to a final configuration in a finite number of cycles from any given initial configuration.

Efficiency of Gathering Algorithms. Given two sets of points $P, W \subseteq \mathbb{R}^2$, we define the *Weber distance* between any point $w \in W$ and P by $wd(P, w) = \sum_{p \in P} |p, w|$, where symbol $|u, v|$ denotes the Euclidean distance between points u and v. A point $\bar{w} \in W$ is the *Weber point* of P w.r.t. W if it minimizes the Weber distance between P and any point $w \in W$, i.e., if $\bar{w} = \text{argmin}_{w \in W} \, wd(P, w)$. The set containing all the Weber points of P w.r.t. W is denoted by $wp(P, W)$. It is well known that if the points in P are not on a line, then the set $wp(P, \mathbb{R}^2)$ contains one point exactly [17]. The Weber points in $wp(P, \mathbb{R}^2)$ might yield a solution for GATHERING. Unfortunately, $wp(P, \mathbb{R}^2)$ is not computable in general – not even with radicals [4]. Instead, if we consider the problem GMP for a given configuration $C = (R, F)$, the Weber points in $wp(R, F)$ can be easily computed (remember that F is finite). Additionally, if we define the *total distance* as the distance traveled by all robots to reach a final configuration, then we can use points in $wp(R, F)$ to measure the efficiency of a gathering algorithm, as stated in the following definition.

Definition 1. *A gathering algorithm for an initial configuration $C = (R, F)$ is* optimal *if it requires the minimum possible total travel distance. Let d be the Weber distance of any point in $wp(R, F)$. Since d is a lower bound for each gathering algorithm, then we say that an algorithm is* exact *if it achieves the gathering with a total distance equal to d.*

Configuration View. Given two distinct points u and v on the plane, let $line(u, v)$ denote the straight line passing through these points and (u, v) (resp. $[u, v]$) denote the open (resp. closed) interval containing all points in this line that lie between u and v. The half-line starting at point u (but excluding the point u) and passing through v is denoted by $hline(u, v)$. Given two lines $line(c, u)$ and $line(c, v)$, we denote by $\sphericalangle(u, c, v)$ the convex angle (i.e., the angle which is at most $180°$) centered in c and with sides $line(c, u)$ and $line(c, v)$.

Given a set P of n distinct points in the plane and an additional point $c \notin P$, let $\bigcup_{p \in P} hline(c, p)$ be the set of all half-lines starting from c and passing through each point in P. The successor of $p \in P$ with respect to c, denoted by $succ(p, c)$, is defined as the point $q \in P$ such that

- either q is the closest point to p on $hline(c, p)$, with $|c, q| > |c, p|$;
- or $hline(c, q)$ is the half-line following $hline(c, p)$ in the order implied by the clockwise direction, and q is the closest point to c on $hline(c, q)$.

Symmetrically, given a point $q \in P$, the predecessor of q with respect to c, denoted by $pred(q, c)$, is the point $p \in P$ such that $succ(p, c) = q$.

Given a configuration $C = (R, F)$, we now use the functions $succ()$ and $pred()$ to define the "view" of points in C. Let $cg(F)$ be the $center\ of\ gravity$ of points in F, and assume there are no points in $U(R) \cup F$ coincident with $cg(F)$. Now, if we take $p \in U(R) \cup F$ and $P = (U(R) \cup F \cup \{cg(F)\}) \setminus \{p\}$, then the function $succ()$ defines the cyclic sequence $V^+(p) = (p_0, p_1, \ldots, p_{n+m-1})$, where $p_0 = cg(F)$ and $p_i = succ(p_{i-1})$[1] for $i \geq 1$. In other words, $V^+(p)$ represents the order in which p views all the points in C starting from $cg(F)$ and turning clockwise according to $succ()$.

From the sequence $V^+(p)$ we directly get the string $\mathcal{V}^+(p)$, that is the $clockwise\ view$ of p, as follows: replace p_i in $V^+(p)$ by the triple α_i, d_i, x_i in $\mathcal{V}^+(p)$, where $\alpha_i = \sphericalangle(p_0, p, p_i)$, $d_i = |p, p_i|$, and $x \in \{\text{"}r\text{"}, \text{"}f\text{"}, \text{"}c\text{"}, \text{"}m\text{"}\}$ according whether p_i is a robot position, a meeting point, the center of gravity of F, or a robot position where a multiplicity occurs, respectively. Similarly, the function $pred()$ allows us to define the $counterclockwise\ view$ of p, denoted by $\mathcal{V}^-(p)$.

The $view\ of\ p$ is defined as $\mathcal{V}(p) = \{\mathcal{V}^+(p), \mathcal{V}^-(p)\}$, and the $view\ of\ the$ $configuration\ C$ is defined as $\mathcal{V}(C) = \bigcup_{p \in U(R) \cup F} \mathcal{V}(p)$. Moreover, defining "$c$" $\leq$ "r" $\leq$ "m" $\leq$ "f" for the third component in the triples used to define $\mathcal{V}^+(p)$ from $V^+(p)$, it is possible to order all the strings in $\mathcal{V}(C)$, and hence defining the $minimum\ view$ of such set.

Notice that $\mathcal{V}()$ has been defined assuming no points in $U(R) \cup F$ coincident with $cg(F)$. If there are $r \in U(R)$ and $f \in F$ coincident with $cg(F)$ (or just one of them), it is possible to define their view as follows. Let p' be the point different from $cg(F)$ and having the minimum view in $\mathcal{V}(C')$, where C' is C without r and f. The view of $p \in \{r, f\}$ is now defined as $V^+(p) = (p_0, p_1, p_2, p_3, \ldots, p_{n+m-1})$, where $p_0 = cg(F)$, $p_1 = r$, $p_2 = f$, $p_3 = p'$, and $p_i = succ(p_{i-1})$ for $i \geq 4$. Then, $\mathcal{V}^+(p)$ is produced from $V^+(p)$ as usual. Finally, two additional concepts about view are needed:

- if $p \in U(R) \cup F$ and $S \subseteq U(R) \cup F$, then $min\_view(p, S)$ says whether p is the point with minimum view in S or not;
- if $f \in F$, then $start(f)$ represents the point(s) in R closest to f but not on it, and having the minimum view in case of ties.

Configuration Automorphisms and Symmetries. Let $\varphi : \mathbb{R}^2 \to \mathbb{R}^2$ a map from points to points in the plane. It is called an $isometry$ or distance preserving

---

[1] If points $r \in U(R)$ and $f \in F$, different from p, are coincident, then points r, f will appear in this order in $V^+(p)$.

if for any $a, b \in \mathbb{R}^2$ one has $|\varphi(a), \varphi(b)| = |a, b|$. Examples of isometries in the plane are *translations*, *rotations* and *reflections*.

An *automorphism* of a configuration $C = (R, F)$ is an isometry from $\mathbb{R}^2$ to itself, that maps robots to robots (i.e., points of R into R) and meeting points to meeting points (i.e., points of F into F). The set of all automorphisms of C forms a group with respect to the composition called *automorphism group* of C and denoted by $\mathrm{Aut}(C)$.

The isometries in $\mathrm{Aut}(C)$ are the identity, rotations, reflections and their compositions. An isometry φ is a rotation if there exists a unique point x such that $\varphi(x) = x$ (and x is called *center of rotation*); it is a reflection if there exists a line ℓ such that $\varphi(x) = x$ for each point $x \in \ell$ (and ℓ is called *axis of symmetry*). Translations are not possible as the sets R and F are finite. Note that the existence of two or more reflections imply a rotation.

If $|\mathrm{Aut}(C)| = 1$, that is C admits only the identity automorphism, then C is said *asymmetric*, otherwise it is said *symmetric* (i.e., C admits rotations or reflections). If C is symmetric due to an isometry φ, a robot cannot distinguish its position at $r \in R$ from $r' = \varphi(r)$. As a consequence, two robots (e.g., one on r and one on $\varphi(r)$) can decide to move simultaneously, as any algorithm is unable to distinguish between them. In such a case, there might be a so called *pending move*, that is one of the robots allowed to move performs its entire Look-Compute-Move cycle while one of the others does not perform the Move phase, i.e. its move is pending. Clearly, all the other robots performing their cycle are not aware whether there is a pending move, that is they cannot deduce the global status from their view. This fact greatly increases the difficulty to devise a gathering algorithm for symmetric configurations.

Given an isometry $\varphi \in \mathrm{Aut}(C)$, the *cyclic subgroup* of order k generated by φ is given by $\{\varphi^0, \varphi^1 = \varphi, \varphi^2 = \varphi \circ \varphi, \ldots, \varphi^{k-1}\}$ where φ^0 is the identity. A reflection ρ generates a cyclic subgroup $H = \{\rho^0, \rho\}$ of order two. The cyclic subgroup generated by a rotation ρ can have any order greater than one. If H is a cyclic subgroup of $\mathrm{Aut}(C)$, the *orbit* of a point $p \in R \cup F$ is $Hp = \{\gamma(p) \mid \gamma \in H\}$. Note that the orbits Hr, for each $r \in R$ form a partition of R. The associated equivalence relation is defined by saying that r and r' are *equivalent* if and only if their orbits are the same, that is $Hr = Hr'$.

Next theorem provides a relationship between isometries and the configuration view.

Theorem 1. *An initial configuration $C = (R, F)$ admits a reflection (rotation, resp.) if and only if there exist two distinct points p and q, belonging both in R or in F, such that $\mathcal{V}^+(p) = \mathcal{V}^-(q)$ ($\mathcal{V}^+(p) = \mathcal{V}^+(q)$, resp.).*

3   Basic Results

In this section we provide some useful properties about Weber points. In the remainder, we use the simple sentence "robot r moves toward a meeting point f" to mean that r *performs a straight move toward f and the final position of*

r *lies on the closed interval* $[r, f]$. We start by observing that it is easy to verify (see also [3]) the following result.

Lemma 1. *Let $C = (R, F)$ be a configuration, $f \in wp(R, F)$, and $r \in R$. If R' represents the robots' positions after r moved toward f, then f is in $wp(R', F)$.*

A corollary of this lemma implies that after the movement of r toward a Weber point f, the set of Weber points is restricted to the meeting points lying on the half-line $hline(r, f)$.

Corollary 1. *Given a configuration $C = (R, F)$, $f \in wp(R, F)$, and $r \in R$. If R' represents the robots' positions after r moved toward f, then all the Weber points in $wp(R', F)$ lie on $hline(r, f)$.*

Proof. Let $f' \in wp(R, F)$. After the move of r:

- if f' lies on $hline(r, f)$, then Lemma 1 implies $f' \in wp(R', F)$;
- if f' does not lie on $hline(r, f)$ then it is easy to see that $wd(R', f') > wd(R', f)$, and hence f' does not belong to $wp(R', F)$. □

We are now interested in estimating how many Weber points are still in $wp(R, F)$ after the move of r toward f. To this end, we pose the following general question: "How many Weber points in $wp(R, F)$ lie on a given line in the plane?"

It is well known that the ellipse is the plane curve consisting of all points p whose sum of distances from two given points p_1 and p_2 (i.e., the *foci*) is a fixed number d. Generalizing, a k-ellipse is the plane curve consisting of all points p whose sum of distances from k given points $p_1, p_2, \ldots, p_k$ is a fixed number. In [15], it is shown that a k-ellipse is a strictly-convex curve, provided the foci p_i are not collinear. This implies that a line intersects a k-ellipse in at most two points. Now, if we apply the notion of k-ellipse to the GMP problem, we easily get that $\sum_{r \in R} |p, r| = d$ is a $|R|$-ellipse consisting of all points p whose sum of distances from all robots is a fixed number d. If we set $d = wd(R, f)$ with $f \in wp(R, F)$, then the equation represents the $|R|$-ellipse containing all the Weber points in $wp(R, F)$. In the remaining, such an ellipse will be denoted by $\mathcal{E}_{R,F}$. The following results characterize $\mathcal{E}_{R,F}$ and, in turn, the set of all Weber points after a robot moved toward one of such points.

If $C = (R, F)$ is a configuration in which points in R are collinear, then the *median segment* of R, denoted by $med(R)$, is the segment $[r_1, r_2]$, where r_1 and r_2 are the median points of R (with $r_1 = r_2$ when $|R|$ is odd).

Lemma 2. *Let $C = (R, F)$ be a configuration.*

- *If points in R are not collinear, then $\mathcal{E}_{R,F}$ is either a single point or a strictly-convex curve with non-empty interior;*
- *If points in R are collinear, then $\mathcal{E}_{R,F}$ is either $med(R)$ or a strictly-convex curve with non-empty interior.*

By using Lemma 2 along with Corollary 1, we get the following result.

Lemma 3. *Let $C = (R, F)$ be a configuration. Assume that a robot $r \in R$ moves toward a point $f \in wp(R, F)$ and this move creates a configuration $C' = (R', F)$. Then:*

- *if $\mathcal{E}_{R,F}$ is a strictly-convex curve with non-empty interior, then $wp(R', F)$ contains one or two Weber points only: one is f and the other (if any) lies on $hline(r, f)$;*
- *if $\mathcal{E}_{R,F}$ is a single point (i.e., $wp(R, F) = \{f\}$), then $wp(R', F)$ contains f only;*
- *if $\mathcal{E}_{R,F}$ is $med(R)$ (i.e., $wp(R, F) = med(R) \cap F$), then $wp(R', F) = med(R') \cap F$.*

The next lemma characterizes the Weber points in case of a particular rotation.

Lemma 4. *Let $C = (R, F)$ be a configuration with robots not collinear. If C admits a rotation whose center $c \in F$, then $wp(R, F) = \{c\}$.*

The next theorem provides a sufficient condition for a configuration to be ungatherable, but we first need the following definition:

Definition 2. *Let $C = (R, F)$ be a configuration. An isometry $\varphi \in Aut(C)$ is called* partitive *on $P \subseteq \mathbb{R}^2$ if the cyclic subgroup H generated by φ has order $k > 1$, and $|Hp| = k$ for each $p \in P$.*

Notice that the identity is not partitive. A reflection ρ with axis of symmetry ℓ generates a cyclic group $H = \{\rho^0, \rho\}$ of order two and is partitive on $\mathbb{R}^2 \setminus \ell$. A rotation ρ is partitive on $\mathbb{R}^2 \setminus \{c\}$, where c is the center of rotation and the cyclic subgroup generated by ρ can have any order greater than one. In the reminder, we say that an isometry φ *fixes* a point p when $\varphi(p) = p$. The following theorem provides us a sufficient condition for establishing when a configuration is ungatherable.

Theorem 2. *Given a configuration $C = (R, F)$, and a subset of points $P \subset \mathbb{R}^2$ with $P \cap R = \emptyset$, if there exists an isometry $\varphi \in Aut(C)$ that is partitive on $\mathbb{R}^2 \setminus P$ and fixes the points of P, then any gathering algorithm can not assure the gathering on a point in $\mathbb{R}^2 \setminus P$.*

The following corollary shows that there exist configurations that are potentially gatherable in GMP, but not by means of *exact* gathering algorithms.

Corollary 2. *Let $C = (R, F)$ be a configuration, and $P \subset \mathbb{R}^2$ with $P \cap (R \cup wp(R, F)) = \emptyset$. If there exists an isometry $\varphi \in Aut(C)$ that is partitive on $\mathbb{R}^2 \setminus P$ and fixes the points of P, then there not exists any exact gathering algorithm for C.*

The following corollary shows that some configurations are ungatherable in GMP.

Corollary 3. *Let $\varphi \in Aut(C)$ be an isometry of a configuration $C = (R, F)$. C is ungatherable if:*

- *φ is a rotation and the center $c \notin R \cup F$;*
- *φ is a reflection with axis ℓ and $\ell \cap (R \cup F) = \emptyset$.*

4   Exact Gathering

In this section we present a distributed algorithm for the problem GMP that assures exact gathering. According to Corollaries 2 and 3, there are initial configurations that cannot be gathered by any exact gathering algorithm. These correspond to configurations C such that:

- C admits a rotation with center c, and there are neither robots nor meeting points on c, or
- C admits a reflection on axis ℓ, and there are neither robots nor Weber points on ℓ.

We denote the above set of configuration by $\mathcal{U}$, and we provide a gathering algorithm that assures exact gathering for all the remaining initial configurations. All the initial configurations processed by the algorithm, along with configurations created during the execution, are partitioned in the following classes:

$\mathcal{S}_1$: any configuration C with one multiplicity;
$\mathcal{S}_2$: any $C = (R, F)$ with $|wp(R, F)| = 1$, and $C \notin \mathcal{S}_1$;
$\mathcal{S}_3$: any $C = (R, F)$ with $cg(F) \in wp(R, F)$, and $C \notin \bigcup_{i=1}^{2} \mathcal{S}_i$;
$\mathcal{S}_4$: any $C = (R, F)$ with all points in R and all points in $wp(R, F)$ lying on a line ℓ, and $C \notin \bigcup_{i=1}^{3} \mathcal{S}_i$;
$\mathcal{S}_5$: any C admitting a rotation, and $C \notin \bigcup_{i=1}^{4} \mathcal{S}_i$;
$\mathcal{S}_6$: any C admitting a reflection with at least one robot and one Weber point on the axis, and $C \notin \bigcup_{i=1}^{5} \mathcal{S}_i$;
$\mathcal{S}_7$: any C admitting a reflection with at least one robot on the axis, and $C \notin \bigcup_{i=1}^{6} \mathcal{S}_i$;
$\mathcal{S}_8$: any C admitting a reflection with at least one Weber point on the axis, and $C \notin \bigcup_{i=1}^{7} \mathcal{S}_i$;
$\mathcal{S}_9$: any asymmetric configuration C, and $C \notin \bigcup_{i=1}^{4} \mathcal{S}_i$;
$\mathcal{S}_0$: $\mathcal{S}_1 \cup \mathcal{S}_2$ (class defined for sake of convenience only).

The main strategy of the algorithm is to select and move robots straightly toward a Weber point f so that, after a certain number of moves, f remains the only Weber point (hence reaching a configuration in class $\mathcal{S}_2$). Once only one Weber point exists, all robots move toward it. According to the global weak multiplicity detection, once a multiplicity is created, robots are no longer able to compute the Weber points accurately. Hence, our strategy assures to create the first multiplicity over f, and once this happens all robots move toward it without creating other multiplicities. Note that, in the initial configuration, it is possible that there was a robot on f. Hence, it is always possible to create a configuration in class $\mathcal{S}_1$ without passing for a configuration in class $\mathcal{S}_2$. Figure 1 shows all transitions among classes defined by the algorithm, and, in particular, it shows that from each class $\mathcal{S}_i$, $i \geq 3$, a configuration in class $\mathcal{S}_1$ or $\mathcal{S}_2$ (i.e., in class $\mathcal{S}_0$) is reached.

Note that configurations in class $\mathcal{S}_1$ are the only non-initial ones. The algorithm is divided into various sub-procedures, each of that designed to process configurations belonging to a given class $\mathcal{S}_i$. Priorities among procedures are implicitly defined by the subscripts in the name of the classes. Since such procedures determine the Compute phase of a robot, after each instruction determining the move of the executing robot, as well as at the end of each procedure, instruction *exit()* must be implicitly considered. Moves are always computed without overtake robots, that is, undesired multiplicities are never created. Moreover, given a configuration $C = (R, F)$, all procedures are invoked after having computed the class $\mathcal{S}_i$ to which C belongs. For this task robots can exploit the multiplicity detection capability (for class $\mathcal{S}_1$), the computation of $wp(R, F)$ and $cg(F)$ (for classes $\mathcal{S}_2$ and $\mathcal{S}_3$), the fact whether there exists a robot whose view contains all robots and Weber points associated with a same angle (for class $\mathcal{S}_4$), and Theorem 1 (for classes $\mathcal{S}_5 - \mathcal{S}_9$). The next theorem provides the correctness proof of our algorithm that is based of the subsequent theorems provided for each possible class.

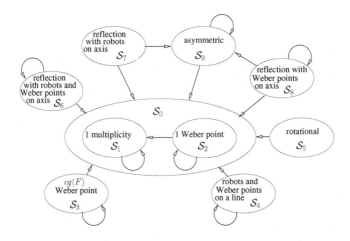

Fig. 1. Schematization of the exact gathering algorithm.

Theorem 3 (correctness). *There exists an exact gathering algorithm for any initial configuration C if and only if $C \notin \mathcal{U}$.*

Proof. ($\Longrightarrow$) We prove this part of the claim by showing that if C belongs to $\mathcal{U}$ then exact gathering cannot be assured. If C admits a rotation with center c, and there are neither robots nor meeting points on c, then C is partitive. The same holds if C admits a reflection on axis ℓ, and there are neither robots nor meeting points on ℓ. Hence, by Corollary 3, C is ungatherable. If C admits a reflection on axis ℓ, and there are neither robots nor Weber points on ℓ, then by Corollary 2, C can be potentially gathered on meeting points lying on ℓ but not in an exact way.

($\Longleftarrow$) If $C \notin \mathcal{U}$, then $C \in \bigcup_{2 \le i \le 9} \mathcal{S}_i$. For each class we define a strategy. The overall behavior of the robots is shown in Fig. 1. Theorems 4–9 (along with arguments in Sect. 4.1) prove that the transitions among classes are those shown in the figure. The only cycles are the self-loops of various classes. However, the corresponding theorems prove that the exit transitions will be used, eventually. Hence, starting from any initial configuration, class $\mathcal{S}_1$ will be reached, eventually, and from there exact gathering can be finalized by letting all robots move toward the unique multiplicity created on a Weber point.   □

4.1   Classes $\mathcal{S}_1$, $\mathcal{S}_2$, and $\mathcal{S}_3$

From classes $\mathcal{S}_1$, $\mathcal{S}_2$, and $\mathcal{S}_3$, robots can move concurrently toward the unique multiplicity, the only Weber point, or $cg(F)$, respectively.

4.2   Class $\mathcal{S}_4$: All Robots and Weber Points on a Line

In this section, we consider the case where all robots and all Weber points lie on a line. After the Look phase, a robot can detect whether the current configuration admits such a property if there exists a robot whose view contains all robots and Weber points associated with a same angle.

Theorem 4. *Given an initial configuration $C = (R, F)$ in class $\mathcal{S}_4$, our exact gathering algorithm leads to a configuration C' in class $\mathcal{S}_0$, eventually.*

Proof. If F admits a reflection with axis ℓ' perpendicular to ℓ, and the intersection between ℓ and ℓ' is $w \in wp(R, F)$, then all robots can move toward w. Note that, as the set F is finite, there cannot be two parallel axis of symmetry induced by F, hence ℓ' is always recognizable (this leads to C' belonging to class $\mathcal{S}_0$, eventually).

If C does not admit an axis of reflection perpendicular to ℓ, then if there are robots in between Weber points, the algorithm makes move the one with minimum view toward a Weber point in any direction. In this way at the subsequent step $|wp(R, F)| = 1$ holds (this leads to class $\mathcal{S}_0$). If there are no robots in between Weber points, then by Lemma 3 all Weber points are in between two robots r_1 and r_2. Let k_1 and k_2 be the number of meeting points that are not Weber points lying on ℓ closer to r_1 and r_2, respectively. Let f_1 and f_2 be the Weber points closest to r_1 and r_2, respectively. If $k_1 > k_2$, then the algorithm makes move r_1 toward f_2. Property $k_1 > k_2$ remains inviolated, and r_1 will be always selected to move until the unique Weber point left is f_2 (this leads to class $\mathcal{S}_0$). If $k_1 = k_2$ and $|r_1, f_1| < |r_2, f_2|$, again r_1 is selected to move toward f_2, and the property remains inviolated until k_1 becomes bigger than k_2. As shown before, still r_1 will be selected to move. If $k_1 = k_2$ and $|r_1, f_1| = |r_2, f_2|$, since the configuration is assumed to be asymmetric, the views of r_1 and r_2 differ. If $\mathcal{V}(r_1) < \mathcal{V}(r_2)$ then again r_1 is selected to move toward f_2, hence reducing its distance to f_1, or increasing k_1 in case it overtakes f_1.

If C admits an axis of reflection ℓ' perpendicular to ℓ then we need to consider three subcases: (i) there are neither robots nor Weber points on ℓ'; (ii) there is a robot r on ℓ'; (iii) there is a Weber point on ℓ'. In case (i), either the configuration is partitive, or there are meeting points on ℓ' that are not Weber points. In either cases, by Corollaries 2 and 3, exact gathering cannot be assured and hence $C \notin \mathcal{S}_4$. In case (ii), by Lemma 3, there are exactly two Weber points on ℓ, at the two sides of r. The algorithm makes move r toward any direction on ℓ, hence obtaining a configuration with a unique Weber point (this leads to class $\mathcal{S}_0$). In case (iii), we are back to the case F admits a reflection that detects a unique Weber point lying on the intersection of ℓ' with ℓ.   □

4.3   Class $\mathcal{S}_5$: Rotations

Theorem 5. *Given an initial configuration $C = (R, F)$ in class $\mathcal{S}_5$, our exact gathering algorithm leads to a configuration C' in class $\mathcal{S}_0$, eventually.*

Proof. Let c be the center of the rotation. Notice that $c \notin F$, otherwise Lemma 4 implies $wp(R, F) = \{c\}$ (and hence C should belong to class $\mathcal{S}_2$, against hypothesis). Notice also that $c \in R$, otherwise from $c \notin R \cup F$ Corollary 3 implies C partitive, and this contradicts C in class $\mathcal{S}_5$ too. Hence $c \in R$ and the robot on c is inside $\mathcal{E}_{R,F}$. The algorithm makes move the robot on c toward an arbitrary point in $wp(R, F)$. By Lemma 3, once the robot has moved, only one Weber point remains. It follows that a configuration C' in class $\mathcal{S}_0$ is created.   □

4.4   Class $\mathcal{S}_6$: Reflections with Robots and Weber Points on the Axis

Theorem 6. *Given an initial configuration $C = (R, F)$ in class $\mathcal{S}_6$, our exact gathering algorithm leads to a configuration C' in class $\mathcal{S}_0$, eventually.*

Proof. Since not all robots and all Weber points are on the axis ℓ, by Lemma 3 there can be at most two Weber points on ℓ.[2]

If there is only one Weber point f on ℓ, the robot on ℓ with minimum view moves toward f. After this movement, only one Weber point remains (thus obtaining a configuration $C' \in \mathcal{S}_0$).

Let us assume there are two Weber points f_1 and f_2. If there are robots in between f_1 and f_2, then the one with minimum view is moved toward either f_1 or f_2, indiscriminately. Once it has moved, only one Weber point is left, and again a configuration $C' \in \mathcal{S}_0$ is obtained.

If there are no robots in between f_1 and f_2, then the closest robot to the two Weber points with minimum view in case of ties, moves toward the farthest Weber point lying on the axis. In this way, it always remains the closest robot to the Weber points on the axis, and once it has overtaken the first Weber point, only one Weber point remains (thus obtaining a configuration $C' \in \mathcal{S}_0$).   □

---

[2] Configurations in class $\mathcal{S}_4$, that is all robots and all Weber points are collinear, have been already addressed.

4.5 Class S_7: Reflections with Robots but No Weber Points on the Axis

Theorem 7. *Given an initial configuration $C = (R, F)$ in class S_7, our exact gathering algorithm leads to a configuration C' in class S_0 or S_9, eventually.*

Proof. Let ℓ be the axis of symmetry, and $\mathcal{E}_{R,F}$ be the R-ellipse where points in $wp(R, F)$ reside.

Let us consider the set of robots $R' = \{p \in R \cap \ell \mid \exists\, f \in wp(R, F) \wedge hline(p, f) \cap wp(R, F) = \{f\}\}$. If R' is non-empty, then select the robot $p \in R'$ with minimum view and move it toward $f \in wp(R, F)$ such that $hline(p, f) \cap wp(R, F) = \{f\}$. After the move, by Lemma 3, f remains the unique Weber point (and hence a configuration in class S_0 is reached). If R' is empty, by Lemma 3, for each robot r on ℓ and for each $f \in wp(R, F)$ we have $|hline(r, f) \cap wp(R, F)| = 2$. It follows that each robot on ℓ is in not inside $\mathcal{E}_{R,F}$, otherwise R' would be not empty.

Let p be a robot on ℓ such that the angle between ℓ and the line passing through p and a $f_p, f'_p \in wp(R, F)$ (say ℓ') is minimum, and denote such an angle as α. Wlog, we assume f_p closer to p than f'_p. Note that there are no Weber points in the arcs of $\mathcal{E}_{R,F}$ from f_p to ℓ and from f'_p to ℓ, otherwise the angle α is not minimum.

We now show that p is the only robot on ℓ.[3] Consider the half-line h of ℓ starting from p and non crossing $\mathcal{E}_{R,F}$. There are no robots p' on h otherwise the angle between ℓ and the line passing through p' and f_p would be less than α. Consider the half-line $h' = \ell \setminus h$ and a point p' on h': either the angle between ℓ and the line passing through p' and f_p is smaller than α, or R' is not empty (i.e., either $hline(p', f_p) \cap wp(R, F) = \{f_p\}$ or $hline(p', f'_p) \cap wp(R, F) = \{f'_p\}$).

The algorithm makes move p toward a point between p and f_p. This point is accurately selected in order to avoid symmetries and to reach a configuration in class S_9.

The new potential symmetry due to the move of p can define ℓ' as an axis of reflection, but in this case the axis was holding also before the movement of p, that is, the configuration was rotational against the hypothesis. Then, according to the definition of isometry, the new potential symmetry must reflect f_p to f'_p, but it cannot be rotational as the $cg(F)$ should lie on ℓ' between f and f'. It remains the case in which there is a new axis of reflection ℓ'' perpendicular to ℓ'. In this case, if there is a robot p' on ℓ' such that $(R \setminus \{p, p'\}, F)$ is symmetric with axis ℓ'', then the algorithm makes move p on ℓ' toward a point at distance d from f_p greater than the distance of p' from f'_p (e.g., $d = (|p, f_p| - |p', f'_p|)/2$). If robot p' does not exists, then the algorithm makes move p toward f_p. In either cases, a configuration in class S_0 is reached. □

---

[3] As a consequence, there is a number odd of robots.

4.6   Class $\mathcal{S}_8$: Reflections with Weber Points but No Robots on the Axis

Theorem 8. *Given an initial configuration $C = (R, F)$ in class $\mathcal{S}_8$, our exact gathering algorithm leads to a configuration C' in class $\mathcal{S}_0$ or $\mathcal{S}_9$, eventually.*

Proof. Consider $f \in wp(R, F)$ on the axis of symmetry ℓ with minimum view. Let $p = start(f)$, and p' be its specular robot wrt ℓ. Before letting move p and p' toward f, the algorithm simulates the move of one robot among them, say p. If the resulting configuration admits two Weber points,[4] one of which is f and another one denoted by f', then the algorithm makes a check. If there is another robot q that appears after p, f and f' on the line where they reside, and such that $|p, f| > |q, f'|$, then the algorithm makes move q and its specular robot q' toward f.

In any case, if both the allowed robots move synchronously, only f remains (i.e., a configuration in class $\mathcal{S}_0$ is reached). If only one robot moves and there are two Weber points left, only one of them has minimum view, hence the configuration is asymmetric (i.e., a configuration in class $\mathcal{S}_9$ is reached). Clearly, since the algorithm has allowed two specular robots to move, it is possible that one of them is pending while the configuration is asymmetric.   □

4.7   Class $\mathcal{S}_9$: Asymmetric

Theorem 9. *Given an initial configuration $C = (R, F)$ in class $\mathcal{S}_9$, our exact gathering algorithm leads to a configuration in class $\mathcal{S}_0$, eventually.*

Proof. We first define what we call *basic strategy* for initial asymmetric configurations (that is, without pending moves). Among all Weber points, let f be the one minimizing $|start(f), f|$, and of minimum view in case of ties. The basic strategy makes move $r = start(f)$ that is unique toward f. By Lemma 3, after the move, there are one or two Weber points only, but still f remains the Weber point with the closest robot, and no ties are possible. If f is the only Weber point left, the obtained configuration C' belongs to class $\mathcal{S}_0$. If there is another Weber point f', then C' cannot admit a symmetry mapping f to f' since $|start(f'), f'| > |r, f|$. Moreover, C' cannot admit a reflection with axis passing through r, f, f', since such a symmetry should have been holding also before the movement. Hence C' belongs to class $\mathcal{S}_9$, and r will be selected again until reaching f. Eventually, either only one Weber point remains or a multiplicity is created (i.e., a configuration in class $\mathcal{S}_0$ is reached).

Actually, the strategy to be applied from asymmetric configurations must take care of possible movements that are pending due to algorithms applied from symmetric configurations. In particular, an asymmetric configuration with two Weber points, say f_1 and f_2, can be obtained from configurations admitting a reflection with only Weber points on the axis (class $\mathcal{S}_7$) or only robots (class $\mathcal{S}_8$). From Theorem 7, it is not possible to obtain the same asymmetric configuration with two

---

[4] By Lemma 3, there cannot be more than two Weber points.

Weber points starting from two different configurations belonging to either $\mathcal{S}_7$ or $\mathcal{S}_8$. In fact, the number of robots of configurations in class $\mathcal{S}_8$ is even, while it is odd for the considered configurations in $\mathcal{S}_7$. Since the move of the basic strategy for asymmetric configurations correctly finalizes the gathering for configurations obtained from class $\mathcal{S}_7$, we only need to design a strategy able to recognize whether the configuration has been potentially obtained from class $\mathcal{S}_8$.

Since robots are oblivious, there is no way to remember whether the initial configuration admitted a symmetry. Then, before applying the basic strategy, the algorithm makes the following check: it looks for the robot p that (i) lies on the line ℓ passing through f_1 and f_2, and (ii) does not lie on $[f_1, f_2]$, and (iii) is closest to the points $\{f_1, f_2\}$. Wlog, let $|p, f_1| < |p, f_2|$.

If p exists and it is unique, the idea is that p has moved before, and there must be another robot p' initially specular to p that might be pending, that is, p' has already performed its Look phase while the configuration was symmetric. As shown by Theorem 8, the same situation can be reached according to two different strategies. Let ℓ_1 (ℓ_2, resp.) be the line passing through $cg(F)$ and f_1 (f_2, resp.). These lines are well defined since $cg(F)$ cannot coincide with f_1 or f_2 as the configurations would belong to $\mathcal{S}_3$. Lines ℓ_1 and ℓ_2 represent the only possible axes of symmetry occurring before the movement of p according to the specified strategies. Let ℓ' (ℓ'', resp.) be the line symmetric to ℓ w.r.t. ℓ_1 (ℓ_2, resp.). If there was a symmetry according to ℓ_2, then the algorithm looks for a robot p'' along the half-line specular to $hline(f_1, p)$ on ℓ'' with minimum $|p'', f_2| > |p, f_2|$. If $(R \setminus \{p, p'\}, F)$ is symmetric, then the algorithm makes move p'' toward f_1.

If the original axis was ℓ_1, then only p' can be pending toward f_1. By allowing p'' to move toward f_1 still maintains f_1 as gathering point and does not create multiplicities out of f_1.

If the original axis was ℓ_2, then only p'' can be pending toward f_2. If p'' is pending, then its move cannot change and after it, the final gathering point will be f_2, otherwise p'' moves toward f_1. In any case, p'' determines the final gathering point assuring exact gathering since all moves are performed toward the point that will be chosen by p'' according to the occurring events. If ℓ_2 couldn't have been a reflection axis, then ℓ_1 is checked as potential reflection axis. The algorithm looks for a robot p' closest to f_1 on the half-line specular to $hline(f_1, p)$. If without considering p and p' the configuration is symmetric, then the algorithm makes move p' toward f_1.

In all cases, only one Weber point remains or a multiplicity is created, i.e., class $\mathcal{S}_0$ is reached.   □

5   Conclusion

We have studied the gathering problem under the Look-Compute-Move cycle model with the global weak multiplicity detection assumption where robots must gather at some predetermined points. A new theory has been devised, and a characterization for pursuing optimal gathering in terms of covered distances has been addressed. We have proposed a distributed algorithm working for any initial configuration where exact gathering has not been proven to be unfeasible. This leaves open the task of designing an optimal algorithm also for those configurations that are potentially gatherable but not in the exact way.

This is the first time this kind of constrained gathering has been addressed. Previous strategies/settings can be now reconsidered with respect to the new twofold objective function that requires to accomplish the gathering task on meeting points while minimizes the overall distances covered by robots.

References

1. Bajaj, C.: The algebraic degree of geometric optimization problems. Discrete Comput. Geom. **3**(1), 177–191 (1988)
2. Bouzid, Z., Das, S., Tixeuil, S.: Gathering of mobile robots tolerating multiple crash faults. In: IEEE 33rd Intenational Conference on Distributed Computing Systems (ICDCS), pp. 337–346 (2013)
3. Cieliebak, M., Flocchini, P., Prencipe, G., Santoro, N.: Distributed computing by mobile robots: gathering. SIAM J. Comput. **41**(4), 829–879 (2012)
4. Cockayne, E.J., Melzak, Z.A.: Euclidean constructibility in graph-minimization problems. Math. Mag. **42**(4), 206–208 (1969)
5. D'Angelo, G., Di Stefano, G., Navarra, A.: Gathering asynchronous and oblivious robots on basic graph topologies under the look-compute-move model. In: Alpern, S., et al. (eds.) Search Theory: A Game Theoretic Perspective, pp. 197–222. Springer, New York (2013)
6. D'Angelo, G., Di Stefano, G., Navarra, A.: Gathering on rings under the look-compute-move model. Distrib. Comput. **27**(4), 255–285 (2014)
7. Di Stefano, G., Navarra, A.: Optimal gathering of oblivious robots in anonymous graphs. In: Moscibroda, T., Rescigno, A.A. (eds.) SIROCCO 2013. LNCS, vol. 8179, pp. 213–224. Springer, Heidelberg (2013)
8. Flocchini, P., Prencipe, G., Santoro, N.: Distributed Computing by Oblivious Mobile Robots. Synthesis Lectures on Distributed Computing Theory, Morgan & Claypool Publishers (2012)
9. Fujinaga, N., Ono, H., Kijima, S., Yamashita, M.: Pattern formation through optimum matching by oblivious CORDA robots. In: Lu, C., Masuzawa, T., Mosbah, M. (eds.) OPODIS 2010. LNCS, vol. 6490, pp. 1–15. Springer, Heidelberg (2010)
10. Klasing, R., Markou, E., Pelc, A.: Gathering asynchronous oblivious mobile robots in a ring. Theor. Comput. Sci. **390**, 27–39 (2008)
11. Kranakis, E., Krizanc, D., Markou, E.: The Mobile Agent Rendezvous Problem in the Ring. Morgan & Claypool, San Rafael (2010)
12. Kupitz, Y., Martini, H.: Geometric aspects of the generalized Fermat-Torricelli problem. Intuitive Geom. Bolyai Soc. Math Stud. **6**, 55–127 (1997)
13. Pelc, A.: Deterministic rendezvous in networks: a comprehensive survey. Networks **59**(3), 331–347 (2012)
14. Prencipe, G.: Impossibility of gathering by a set of autonomous mobile robots. Theoret. Comput. Sci. **384**, 222–231 (2007)
15. Sekino, J.: n-ellipses and the minimum distance sum problem. Amer. Math. Monthly **106**(3), 193–202 (1999)
16. Suzuki, I., Yamashita, M.: Distributed anonymous mobile robots: Formation of geometric patterns. SIAM J. Comput. **28**(4), 1347–1363 (1999)
17. Weiszfeld, E.: Sur le point pour lequel la somme des distances de n points donnés est minimum. Tohoku Math. **43**, 355–386 (1936)

Algorithms and Data Structures
on Graphs

Fast Rendezvous with Advice

Avery Miller$^{(\boxtimes)}$ and Andrzej Pelc

Université du Québec en Outaouais, Gatineau, Canada
`a4miller@cs.toronto.edu, pelc@uqo.ca`

Abstract. Two mobile agents, starting from different nodes of an n-node network at possibly different times, have to meet at the same node. This problem is known as *rendezvous*. Agents move in synchronous rounds using a deterministic algorithm. In each round, an agent decides to either remain idle or to move to one of the adjacent nodes. Each agent has a distinct integer label from the set $\{1, \ldots, L\}$, which it can use in the execution of the algorithm, but it does not know the label of the other agent.

The main efficiency measure of a rendezvous algorithm's performance is its *time*, i.e., the number of rounds from the start of the later agent until the meeting. If D is the distance between the initial positions of the agents, then $\Omega(D)$ is an obvious lower bound on the time of rendezvous. However, if each agent has no initial knowledge other than its label, time $O(D)$ is usually impossible to achieve. We study the minimum amount of information that has to be available *a priori* to the agents to achieve rendezvous in optimal time $\Theta(D)$. Following the standard paradigm of *algorithms with advice*, this information is provided to the agents at the start by an oracle knowing the entire instance of the problem, i.e., the network, the starting positions of the agents, their wake-up rounds, and both of their labels. The oracle helps the agents by providing them with the *same* binary string called *advice*, which can be used by the agents during their navigation. The length of this string is called the *size of advice*. Our goal is to find the smallest size of advice which enables the agents to meet in time $\Theta(D)$. We show that this optimal size of advice is $\Theta(D \log(n/D) + \log \log L)$. The upper bound is proved by constructing an advice string of this size, and providing a natural rendezvous algorithm using this advice that works in time $\Theta(D)$ for all networks. The matching lower bound, which is the main contribution of this paper, is proved by exhibiting classes of networks for which it is impossible to achieve rendezvous in time $\Theta(D)$ with smaller advice.

Keywords: Rendezvous · Advice · Deterministic distributed algorithm · Mobile agent · Time

A. Pelc—Partially supported by NSERC discovery grant and by the Research Chair in Distributed Computing at the Université du Québec en Outaouais.

© Springer-Verlag Berlin Heidelberg 2015
J. Gao et al. (Eds.): ALGOSENSORS 2014, LNCS 8847, pp. 75–87, 2015.
DOI: 10.1007/978-3-662-46018-4_5

1   Introduction

1.1   Background

Two mobile agents, starting from different nodes of a network, have to meet at the same node in the same time. This distributed task is known as *rendezvous* and has received a lot of attention in the literature. Agents can be any mobile autonomous entities. They might represent human-made objects, such as software agents in computer networks or mobile robots navigating in a network of corridors in a building or a mine. They might also be natural, such as animals seeking a mate, or people who want to meet in an unknown city whose streets form a network. The purpose of meeting in the case of software agents or mobile robots might be the exchange of data previously collected by the agents or samples collected by the robots. It may also be the coordination of future network maintenance tasks, for example checking functionality of websites or of sensors forming a network, or decontaminating corridors of a mine.

1.2   Model and Problem Description

The network is modeled as an undirected connected graph with n unlabeled nodes. We seek deterministic rendezvous algorithms that do not rely on the agents perceiving node identifiers, and therefore can work in anonymous graphs as well (cf. [1]). The reason for designing such algorithms is that, even when nodes of the network have distinct identifiers, agents may be unable to perceive them because of limited sensory capabilities (e.g., a mobile robot may be unable to read signs at corridor crossings), or nodes may be unwilling to reveal their identifiers to software agents, e.g., due to security or privacy reasons. From a methodological standpoint, if nodes had distinct identifiers visible to the agents, the agents could explore the graph and meet at the node with smallest ID. In this case, rendezvous reduces to graph exploration.

On the other hand, we assume that, at each node v, each edge incident to v has a distinct *port number* from the set $\{0, \dots, d-1\}$, where d is the degree of v. These port numbers are fixed and visible to the agents. Port numbering is *local* to each node, i.e., we do not assume any relation between port numbers at the two endpoints of an edge. Note that in the absence of port numbers, edges incident to a node would be undistinguishable for agents and thus rendezvous would be often impossible, as an agent may always miss some particular edge incident to the current node, and this edge could be a bridge to the part of the graph where the other agent started. The previously mentioned security and privacy reasons for not revealing node identifiers to software agents are irrelevant in the case of port numbers. If the graph models a system of corridors of a mine or a building, port numbers can be made implicit, e.g., by marking one edge at each intersection (using a simple mark legible even by a mobile robot with very limited vision), considering it as corresponding to port 0, and all other port numbers increasing clockwise.

Agents are initially located at different nodes of the graph and traverse its edges in synchronous rounds. They cannot mark visited nodes or traversed edges in any way, and they cannot communicate before meeting. The adversary wakes up each of the agents, possibly in different rounds. Each agent starts executing the algorithm in the round of its wake-up. It has a clock that ticks at each round and starts at the wake-up round of the agent. In each round, each agent either remains at the current node, or chooses a port in order to move to one of the adjacent nodes. When an agent enters a node, it learns the node's degree and the port number of entry. When agents cross each other on an edge while traversing it simultaneously in different directions, they do not notice this fact.

Each agent has a distinct integer label from a fixed *label space* $\{1, \ldots, L\}$, which it can use as a parameter in the same deterministic algorithm that both agents execute. It does not know the label nor the starting round of the other agent. Since we study deterministic rendezvous, the absence of distinct labels would preclude the possibility of meeting in highly symmetric graphs, such as rings or tori, for which there exist non-trivial port-preserving automorphisms. Indeed, in such graphs, identical agents starting simultaneously and executing the same deterministic algorithm can never meet, since they will keep the same positive distance in every round. Hence, assigning different labels to agents is the only way to break symmetry, as is needed to meet in every graph using a deterministic algorithm. On the other hand, if agents knew each other's identities, then the smaller-labelled agent could stay inert, while the other agent would try to find it. In this case rendezvous reduces to graph exploration. Assuming such knowledge, however, is unrealistic, as agents are often created independently, and they know nothing about each other prior to meeting.

The rendezvous is defined as both agents being at the same node in the same round. The main efficiency measure of a rendezvous algorithm's performance is its *time*, i.e., the number of rounds from the start of the later agent until the meeting. If D is the distance between the initial positions of the agents, then $\Omega(D)$ is an obvious lower bound on the time of rendezvous. However, if the agents have no additional knowledge, time $O(D)$ is usually impossible to achieve. This is due to two reasons. First, without any knowledge about the graph, even the easier task of *treasure hunt* [21], in which a single agent must find a target (treasure) hidden at an unknown node of the graph, takes asymptotically larger time in the worst case. Treasure hunt is equivalent to a special case of rendezvous where one of the agents is inert. In the worst case, this takes as much time as graph exploration, i.e., having a single agent visit all nodes. Second, even when the graph is so simple that navigation of the agents is not a problem, breaking symmetry between the agents, which is often necessary to achieve a meeting, may take time larger than D. Indeed, even in the two-node graph, where $D = 1$, rendezvous requires time $\Omega(\log L)$ [8].

We study the amount of information that has to be given *a priori* to the agents to achieve rendezvous in optimal time $\Theta(D)$. Following the paradigm of *algorithms with advice* [5,7,11–15,17–19,22], this information is provided to the agents at the start, by an oracle knowing the entire instance of the problem, i.e., the graph, the starting positions of the agents, their wake-up rounds, and both of their labels. The oracle helps the agents by providing them with the *same* binary string called *advice*, which can be used by each agent, together with its own label, during the execution of the algorithm. The length of this string is called the *size of advice*. Our goal is to find the smallest size of advice (up to constant factors) which enables the agents to meet in time $\Theta(D)$. In other words we want to answer the question:

What is the minimum information that permits the fastest possible rendezvous? where both "minimum" and "fastest" are meant up to multiplicative constants.

Notice that, since the advice given to both agents is identical, it could not help break symmetry if agents did not have distinct labels. Hence, even with large advice, the absence of distinct labels would preclude rendezvous in highly symmetric networks, as argued above. Using the framework of advice permits us to quantify the amount of information needed for an efficient solution of a given network problem (in our case, rendezvous), regardless of the type of information that is provided.

1.3   Our Results

For agents with labels from the set $\{1, \ldots, L\}$, we show that, in order to meet in optimal time $\Theta(D)$ in n-node networks, the minimum size of advice that has to be provided to the agents is $\Theta(D \log(n/D) + \log \log L)$. The upper bound is proved by constructing an advice string of this size, and providing a natural rendezvous algorithm using this advice that works in time $\Theta(D)$ for all networks. The matching lower bound, which is the main contribution of this paper, is proved by exhibiting classes of networks for which it is impossible to achieve rendezvous in time $\Theta(D)$ with smaller advice.

Our algorithm works for arbitrary starting times of the agents, and our lower bound is valid even for simultaneous start. As far as the memory of the agents is concerned, our algorithm has very modest requirements: an agent must only be able to store the advice and its own label. Hence memory of size $\Theta(D \log(n/D) + \log L)$ is sufficient. On the other hand, our lower bound on the size of advice holds even for agents with unlimited memory.

Omitted proofs will appear in the full version of the paper.

1.4   Related Work

The problem of rendezvous has been studied both under randomized and deterministic scenarios. A survey of randomized rendezvous in various models can be found in [1]. Deterministic rendezvous in networks has been surveyed in [20]. Several authors considered rendezvous in the plane [2,4].

For the deterministic setting, many authors studied the feasibility and time complexity of rendezvous. Most relevant to our work are the results about deterministic rendezvous in arbitrary graphs, when the two agents cannot mark nodes, but have unique labels [8,21]. In [8], the authors present a rendezvous algorithm whose running time is polynomial in the size of the graph, in the length of the shorter label and in the delay between the starting times of the agents. In [21], rendezvous time is polynomial in the first two of these parameters and independent of the delay.

Apart from the synchronous model used in this paper, several authors investigated asynchronous rendezvous in the plane [4] and in network environments [3,6,9]. In the latter scenario, the agent chooses the edge to traverse, but the adversary controls the speed of the agent. Under this assumption, rendezvous at a node cannot be guaranteed even in very simple graphs. Hence the rendezvous requirement is relaxed to permit the agents to meet inside an edge.

Providing nodes or agents with arbitrary kinds of information that can be used to perform network tasks more efficiently has been proposed in [5,7,11–15,17–19,22]. This approach was referred to as *algorithms with advice*. Advice is given either to nodes of the network or to mobile agents performing some network task. In the first case, instead of advice, the term *informative labeling schemes* is sometimes used. Several authors studied the minimum size of advice required to solve network problems in an efficient way.

In [12] the authors compared the minimum size of advice required to solve two information dissemination problems using a linear number of messages. In [14] it was shown that advice of constant size given to the nodes enables the distributed construction of a minimum spanning tree in logarithmic time. In [11] the advice paradigm was used for online problems. In the case of [19] the issue was not efficiency but feasibility: it was shown that $\Theta(n \log n)$ is the minimum size of advice required to perform monotone connected graph clearing. In [7] the task of drawing an isomorphic map was executed by an agent in a graph and the problem was to determine the minimum advice that has to be given to the agent for the task to be feasible.

Among the papers using the paradigm of advice, [5,13] are closest to the present work, as they both concern the task of graph exploration by an agent. In [5] the authors investigated the minimum size of advice that has to be given to unlabeled nodes (and not to the agent) to permit graph exploration by an agent modeled as a k-state automaton. In [13] the authors established the size of advice that has to be given to an agent in order to explore trees while obtaining competitive ratio better than 2. To the best of our knowledge, rendezvous with advice has never been studied before.

2   The Advice and the Algorithm

Consider any n node graph, and suppose that the distance between the initial positions of the agents is D. In this section, we construct an advice string of length $O(D \log(n/D) + \log \log L)$ and a rendezvous algorithm which achieves time D using this advice. We first describe the advice string. Let G be the underlying graph and let ℓ_1 and ℓ_2 be the distinct labels of the agents, both belonging to the label space $\{1, \ldots, L\}$. Call the agent with label ℓ_1 the first agent and the agent with label ℓ_2 the second agent. Let x be the smallest index such that the binary representations of ℓ_1 and ℓ_2 differ on the xth bit. Without loss of generality assume that the xth bit is 0 in the binary representation of ℓ_1 and 1 in the binary representation of ℓ_2.

Let P be a fixed shortest path in G between the initial positions u and v of the agents. The path P induces two sequences of ports of length D: the sequence π' of consecutive ports to be taken at each node of path P to get from u to v, and the sequence π'' of consecutive ports to be taken at each node of path P to get from v to u. Let $\pi \in \{\pi', \pi''\}$ be the sequence corresponding to the direction from the initial position of the second agent to the initial position of the first agent. Denote $\pi = (p_1, \ldots, p_D)$. Let A_i, for $i = 1, \ldots, D$, be the binary representation of the integer p_i. Additionally, let A_0 be the binary representation of the integer x. The binary strings $(A_0, \ldots, A_D)$ will be called substrings.

The sequence of substrings $(A_0, \ldots, A_D)$ is encoded into a single advice string to pass to the algorithm. More specifically, the sequence is encoded by doubling each digit in each substring and putting 01 between substrings. This permits the agent to unambiguously decode the original sequence. Denote by $Concat(A_0, \ldots, A_D)$ this encoding and let $Decode$ be the inverse (decoding) function, i.e., $Decode(Concat(A_0, \ldots, A_D)) = (A_0, \ldots, A_D)$. As an example, $Concat((01), (00)) = (0011010000)$. Note that the encoding increases the total number of advice bits by a constant factor. The advice string given to the agents is $\mathcal{A} = Concat(A_0, \ldots, A_D)$.

The idea of the Algorithm **Fast Rendezvous** using the advice string $\mathcal{A}$ is the following. Each agent decodes the sequence $(A_0, \ldots, A_D)$ from the string $\mathcal{A}$. Then each agent looks at the xth bit of its label, where x is the integer represented by A_0. If this bit is 0, the agent stays inert at its initial position, otherwise it takes the consecutive ports $p_1, \ldots, p_D$, where p_i, for $i = 1, \ldots, D$, is the integer with binary representation A_i. After these D moves, the agent meets the other agent at the latter's initial position. See the pseudocode below.

Algorithm Fast Rendezvous

Input: advice string $\mathcal{A}$, label ℓ.

$(A_0, \ldots, A_D) := Decode(\mathcal{A})$
$x :=$ the integer with binary representation A_0.
if the xth bit of ℓ is 1 **then**
      for $i = 1$ **to** D **do**
            $p_i :=$ the integer with binary representation A_i
            take port p_i
stop.

Theorem 1. *Let G be an n-node graph with two agents initially situated at distance D from one another. Algorithm* **Fast Rendezvous** *achieves rendezvous in time D, using advice of size $O(D \log(n/D) + \log \log L)$.*

3   The Lower Bound

In this section, we prove a lower bound on the size of advice permitting rendezvous in optimal time $O(D)$, where D is the initial distance between the agents. This lower bound will match the upper bound established in Theorem 1, which says that, for an arbitrary n-node graph, rendezvous can be achieved in time $O(D)$ using advice of size $O(D \log(n/D) + \log \log L)$. In order to prove that this size of advice cannot be improved in general, we present two classes of graphs: one that requires advice $\Omega(D \log(n/D))$ and another that requires advice $\Omega(\log \log L)$ to achieve optimal time of rendezvous. To make the lower bound even stronger, we show that it holds even in the scenario where agents start simultaneously.

The $\Omega(D \log(n/D))$ lower bound will be first proved for the simpler problem of treasure hunt. Recall that in this task, a single agent must find a stationary target (treasure) hidden at an unknown node of the graph at distance D from the initial position of the agent. We then show how to derive the same lower bound on the size of advice for the rendezvous problem.

The following technical lemma gives a construction of a graph which will provide the core of our argument for the $\Omega(D \log(n/D))$ lower bound.

Lemma 1. *Let n and D be positive integers such that $D \leq n/2$. Consider any treasure-hunting algorithm A that takes Dz bits of advice. For any fixed even integer $k \in \{2, \ldots, n-1\}$ and every integer $\ell \in \{1, \ldots, \min\{\lfloor \frac{D}{2} \rfloor, \lfloor \frac{n-1}{k} \rfloor\}\}$, there exists a graph of size $k\ell + 1$, an initial position of the agent in this graph, and a location of the treasure at distance 2ℓ from this initial position, for which algorithm A uses $\Omega(\frac{\ell k^2}{2^{2z}})$ rounds.*

Proof. We define a class of graphs $\mathcal{G}(k, \ell)$ such that each graph in $\mathcal{G}(k, \ell)$ has $k\ell + 1$ nodes. We will prove that there is a non-empty subset B of $\mathcal{G}(k, \ell)$ such that, on each graph in B, algorithm A uses $\Omega(\frac{\ell k^2}{2^{2z}})$ rounds to complete treasure hunt, for some initial position of the agent and a location of the treasure at distance 2ℓ from this location.

Each graph G in the class consists of ℓ copies of a k-clique H (with a port numbering to be described shortly), which are chained together in a special way. We will refer to these cliques as $H_1, \ldots, H_\ell$.

Let $v_1, \ldots, v_k$ denote the nodes of H. It should be stressed that names of nodes in cliques are for the convenience of the description only, and they are not visible to the agent. We choose an arbitrary edge-colouring of H using the colours $\{0, \ldots, k-2\}$, which is always possible for cliques of even size [16]. For an arbitrary edge e in H, let $c(e)$ denote the colour assigned to e. The port numbers of H are simply the edge colours, i.e., for any edge $\{u, v\}$, the port numbers corresponding to this edge at u and v are both equal to $c(\{u, v\})$.

Each graph $G \in \mathcal{G}(k, \ell)$ is obtained by chaining together the copies $H_1, \ldots, H_\ell$ of the clique H in the following way. We will call node v_1 in clique H_i the *gate* g_i of H_i. The initial position of the agent is g_1. Each gate g_i, for $i > 1$, is placed on (i.e., subdivides) one of the edges of clique H_{i-1} not incident to g_{i-1}. We denote this edge by e_{i-1}. Finally, an additional *treasure node* $g_{\ell+1}$ is placed on (i.e., subdivides) one of the edges of clique H_ℓ not incident to g_ℓ, and this edge is denoted by e_ℓ. Hence g_1 has degree $k-1$, each g_i, for $1 < i \leq \ell$, has degree $k+1$, and $g_{\ell+1}$ has degree 2, cf. Fig. 1(a). Note that, since g_i, for $i > 1$, subdivides an edge that is not incident to g_{i-1}, we have $D = 2\ell$. Port numbering of graph G is the following. Port numbers in each clique H_i are unchanged, the new port numbers at each node g_i, for $1 < i \leq \ell$, are $k-1$ and k, with $k-1$ corresponding to the edge whose other endpoint has smaller index, and the new port numbers at node $g_{\ell+1}$ are 0 and 1, with 0 corresponding to the edge whose other endpoint has smaller index, cf. Fig. 1(b). All graphs in the class $\mathcal{G}(k, \ell)$ are isomorphic and differ only by port numbering. Note that each graph in $\mathcal{G}(k, \ell)$ is uniquely identified by the sequence of edges $(e_1, \ldots, e_\ell)$. Therefore, the number of graphs in $\mathcal{G}(k, \ell)$ is $N = ((k-1)(k-2)/2)^\ell$.

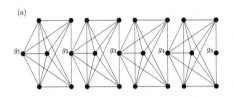
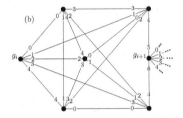

Fig. 1. (a) A graph in $\mathcal{G}(6, 4)$ (b) Port numbering of each clique H_i, for $i < \ell$, with gate g_{i+1} inserted

Notice that an agent navigating in a graph $G \in \mathcal{G}(k, \ell)$ always knows when it arrives at a gate g_i, for $1 < i \leq \ell$, because these are the only nodes of degree $k+1$. An agent's walk is *normal*, if the agent visits each gate g_i, for $1 < i \leq \ell$, exactly once (i.e., never exits a gate by port $k-1$ or k). It is enough to prove our lower bound on the time of treasure hunt only for algorithms where the agent always performs a normal walk. Indeed, for any walk, there exists a normal walk using at most the same time. From now on we restrict attention to such algorithms.

We prove our lower bound on the class of graphs $\mathcal{G}(k, \ell)$. The idea is that, in order to find the treasure node, the agent must visit each of the nodes $g_1, \ldots, g_{\ell+1}$. To get from g_i to g_{i+1}, the agent must find the edge e_i of H_i that the node g_{i+1} subdivides. With little advice, this amounts to searching many edges of the clique H_i, and hence increases time.

For any graph G, the agent is given some advice string S and executes its treasure-hunting algorithm A. With Dz bits of advice, there exists a set B of at least $\frac{N}{2^{Dz}}$ graphs for which the agent is given the same advice string. Next, we provide an upper bound on the number of graphs in B. By comparing this upper bound with $\frac{N}{2^{Dz}}$, we will get the desired lower bound on the number of rounds needed to find the treasure.

Let T be the maximum running time of algorithm A on graphs of class $\mathcal{G}(k, \ell)$. Let τ be the function that maps each graph from $B \subseteq \mathcal{G}(k, \ell)$ to an ℓ-tuple $(t_1, \ldots, t_\ell)$, where, for each $i \in \{1, \ldots, \ell\}$, t_i is the number of edge traversals performed by the agent in clique H_i. This function is well-defined since we consider only deterministic algorithms. The following result shows that this function is injective.

Claim 1. For any two graphs $G \neq G'$ in the set B, we have $\tau(G) \neq \tau(G')$.

By Claim 1, the number of graphs in B is bounded above by the size of the range of τ. Consider an arbitrary ℓ-tuple $(t_1, \ldots, t_\ell)$ in the range of τ. By the definition of $\mathcal{G}(k, \ell)$, for each $i \in \{1, \ldots, \ell\}$, the agent must traverse at least two edges to get from g_i to g_{i+1}. Further, T is an upper bound on the number of edge traversals performed in any execution of the algorithm. Therefore, the size of the range of τ is bounded above by the number of integer-valued ℓ-tuples with positive terms whose sum is at most T. Clearly, this is bounded above by the number of real-valued ℓ-tuples with non-negative terms whose sum is at most T, i.e., by the size of the simplex $\Delta_\ell = \{(t_1, \ldots, t_\ell) \in \mathbb{R}^\ell \mid \sum_{i=1}^{\ell} t_i = T \text{ and } 0 \leq t_i \leq T \text{ for all } i\}$.

From [10], the volume of Δ_ℓ is equal to $T^\ell/\ell!$. Thus, we have shown that the size of B is bounded above by $T^\ell/\ell!$. Comparing this to our lower bound $\frac{N}{2^{Dz}}$ on the size of B, we get

$$T \geq \sqrt[\ell]{\ell! \frac{N}{2^{Dz}}} \geq \sqrt[\ell]{\ell! \frac{((k-1)(k-2)/2)^\ell}{2^{Dz}}} \geq \sqrt[\ell]{\ell!} \frac{(k-2)^2/2}{2^{Dz/\ell}} = \sqrt[\ell]{\ell!} \frac{(k-2)^2/2}{2^{2z}}.$$

By Stirling's formula we have $\ell! \geq \sqrt{\ell}(\ell/e)^\ell$, for sufficiently large ℓ. Hence $\sqrt[\ell]{\ell!} \geq \ell^{1/(2\ell)} \cdot (\ell/e)$. Since the first factor converges to 1 as ℓ grows, we have $\sqrt[\ell]{\ell!} \in \Omega(\ell)$. Hence the above bound on T implies $T \in \Omega(\frac{\ell k^2}{2^{2z}})$. $\qquad\square$

The following theorem follows from Lemma 1 by considering two cases, when $D \in o(n)$ and when $D \in \Omega(n)$, and choosing appropriate values of k and ℓ to apply the lemma in each case.

Theorem 2. *Let n and D be positive integers such that $D \leq n/2$. If an algorithm A solves treasure hunting in $O(D)$ rounds whenever the treasure is at distance D from the initial position of the agent, then there exists an n-node graph G with treasure at this distance such that A requires $\Omega(D \log(n/D))$ bits of advice.*

We can then deduce a lower bound on the size of advice for rendezvous (even with simultaneous start) from the lower bound for treasure hunt.

Corollary 1. *Let $D' \leq n'$ be positive integers. There exist $n \in \Theta(n')$ and $D \in \Theta(D')$ such that if an algorithm A solves rendezvous in time $O(D)$ in n-node graphs whenever the initial distance between the agents is D, then there exists an n-node graph for which A requires $\Omega(D \log(n/D))$ bits of advice.*

The second part of our lower bound on the size of advice, i.e., the lower bound $\Omega(\log \log L)$, will be proved on the class of oriented rings. A ring is *oriented* if every edge has port labels 0 and 1 at the two end-points. Such a port labeling induces orientation of the ring: at each node, we will say that taking port 0 is going clockwise and taking port 1 is going counterclockwise. We assume that agents operate in an oriented ring of size n. In order to make the lower bound as strong as possible, we prove it even for simultaneous start of the agents.

Theorem 3. *Let $D' \leq n'$ be positive integers. Consider any algorithm A that solves rendezvous for agents with labels from the set $\{1, \ldots, , L\}$. There exist $n \in \Theta(n')$ and $D \in \Theta(D')$ such that if A uses time $O(D)$ in the n-node oriented ring whenever the initial distance between the agents is D, then the required size of advice is $\Omega(\log \log L)$.*

Proof. Assume that S is the advice string given to the agents. Consider an agent with label $x \in \{1, \ldots, L\}$ executing algorithm A using advice S. The actions of the agent in consecutive rounds until rendezvous are specified by a *behaviour vector* V_x. In particular, V_x is a sequence with terms from $\{-1, 0, 1\}$ that specifies, for each round i, whether agent x moves clockwise (denoted by -1), remains idle (denoted by 0), or moves counter-clockwise (denoted by 1). Note that an agent's behaviour vector is independent of its starting position, since all nodes of the ring look the same to the agent. This behaviour vector depends exclusively on the label of the agent and on the advice string S.

Let $D = 3D'$, $m = n' - (n' \bmod D')$ and $n = \max(m, 6D')$. Hence $n \in \Theta(n')$, $D \in \Theta(D')$, D' divides n, and $n \geq 2D$. As the initial positions of the agents, fix any nodes v and w of the n-node oriented ring, where w is at clockwise distance D from v. Since $n \geq 2D$, agents are at distance D in the ring. Partition the nodes of the ring into r consecutive blocks $B_0, B_1, \ldots, B_{r-1}$ of size D', starting clockwise from node v. Hence the initial positions v and w of the agents are the clockwise-first nodes of block B_0 and block B_3, respectively. Since agents start simultaneously, we have the notion of global round numbers counted since their start. Partition all rounds $1, 2, \ldots$ into consecutive *time segments* of length D'. Hence, during any time segment, an agent can be located in at most two (neighbouring) blocks.

Fix a behaviour vector V_x of an agent with label x. We define its *meta-behaviour vector* as a sequence M_x with terms from $\{-1, 0, 1\}$ as follows. Suppose that the agent is in block B_j in the first round of the i-th segment. The i-th term of M_x is $z \in \{-1, 0, 1\}$, if, in the first round of the $(i+1)$-th time segment, the agent is in the block B_{j+z}, where index addition is modulo r. Since the initial position of an agent is the clockwise-first node of a block, for a fixed behaviour vector of an agent its meta-behaviour vector is well defined.

Suppose that algorithm A takes at most cD rounds, for some constant c. This corresponds to d time segments for some constant $d \leq 3c$. Hence, all meta-behaviour vectors describing the actions of agents before the meeting are of length d (shorter meta-behaviour vectors can be padded by zeroes at the end.) Let $\mathcal{B}$ be the set of sequences of length d with terms in $\{-1, 0, 1\}$. Sequences from $\mathcal{B}$ represent possible meta-behaviour vectors of the agents. $\mathcal{B}$ has 3^d elements.

Since the initial positions of the agents are in blocks that are separated by two other blocks, agents with the same meta-behaviour vectors must be in different blocks in every round, and hence they can never meet. Indeed, in the first round of every time segment they must be in blocks separated by two other blocks, and during any time segment, an agent can either stay in the same block or get to an adjacent block.

Suppose that the number of bits of advice is at most $\frac{1}{2} \log \log L$. It follows that the set $\mathcal{A}$ of advice strings is of size at most $\sqrt{\log L}$. For any label $x \in \{1, \ldots, L\}$, let Φ_x be the function from $\mathcal{A}$ to $\mathcal{B}$, whose value on an advice string $S \in \mathcal{A}$ is the meta-behaviour vector of the agent with label x when given the advice string S. Functions Φ_x are well-defined, as the meta-behaviour vector of an agent whose initial position is the clockwise-first node of a block depends only on its behaviour vector, which in turn depends only on the agent's label and on the advice string.

If the set $\mathcal{B}^{\mathcal{A}}$ of all functions from $\mathcal{A}$ to $\mathcal{B}$ had fewer elements than L, then there would exist two distinct labels x_1 and x_2 of agents such that, for any advice string S, these agents would have an identical meta-behaviour vector. As observed above, these agents could never meet. This implies $(3^d)^{\sqrt{\log L}} \geq |\mathcal{B}^{\mathcal{A}}| \geq L$. Thus $d \log 3 \geq \sqrt{\log L}$, which contradicts the fact that d is a constant.

Hence the size of advice must be larger than $\frac{1}{2} \log \log L \in \Omega(\log \log L)$.   □

Corollary 1 and Theorem 3 imply:

Theorem 4. *Let $D' \leq n'$ be positive integers. Consider any algorithm A that solves rendezvous for agents with labels from the set $\{1, \ldots, , L\}$. There exist $n \in \Theta(n')$ and $D \in \Theta(D')$ such that, if A takes time $O(D)$ in all n-node graphs whenever the initial distance between agents is D, then the required size of advice is $\Omega(D \log(n/D) + \log \log L)$.*

Theorems 1 and 4 imply the following corollary which is our main result.

Corollary 2. *The minimum size of advice sufficient to accomplish rendezvous of agents with labels from the set $\{1, \ldots, L\}$ in all n-node graphs in time $O(D)$, whenever the initial distance between agents is D, is $\Theta((D \log(n/D) + \log \log L))$.*

4   Conclusion

We established that $\Theta(D \log(n/D) + \log \log L)$ is the minimum amount of information (advice) that agents must have in order to meet in optimal time $\Theta(D)$, where D is the initial distance between them. It should be noted that the two summands in this optimal size of advice have very different roles. On one hand, $\Theta(D \log(n/D))$ bits of advice are necessary and sufficient to accomplish, in $O(D)$ time, the easier task of treasure hunt in n-node networks, where a single agent must find a target (treasure) hidden at an unknown node of the network at distance D from its initial position. This task is equivalent to a special case of rendezvous where one of the agents is inert. On the other hand, for agents whose labels are drawn from a label space of size L, $\Theta(\log \log L)$ bits of advice are needed to break symmetry quickly enough in order to solve rendezvous in time $O(D)$, and hence, are necessary to meet in optimal time $\Theta(D)$, even in constant-size networks. It should be stressed that the first summand in $O(D \log(n/D) + \log \log L)$ is usually larger than the second. Indeed, only when L is very large with respect to n and D does the second summand dominate. This means that "in most cases" the easier task of solving treasure hunt in optimal time is as demanding, in terms of advice, as the harder task of solving rendezvous in optimal time.

In this paper, we assumed that the advice given to both agents is identical. How does the result change when each agent can get different advice? It is clear that giving only one bit of advice, 0 to one agent and 1 to the other, breaks symmetry between them, e.g., the algorithm can make the agent that received bit 0 stay inert. Thus, if advice can be different, one bit of advice reduces rendezvous to treasure hunt. The opposite reduction is straightforward. Hence it follows from our results that $\Theta(D \log(n/D))$ bits of advice are necessary and sufficient to accomplish rendezvous in optimal time $\Theta(D)$ in n-node networks, if advice can be different. This holds regardless of the label space and is, in fact, also true for anonymous (identical) agents.

References

1. Alpern, S., Gal, S.: The Theory of Search Games and Rendezvous. International Series in Operations research and Management Science. Springer, New York (2002)
2. Anderson, E., Fekete, S.: Asymmetric rendezvous on the plane. In: Proceedings of the 14th Annual ACM Symposium on Computational Geometry, pp. 365–373 (1998)
3. Bampas, E., Czyzowicz, J., Gąsieniec, L., Ilcinkas, D., Labourel, A.: Almost optimal asynchronous rendezvous in infinite multidimensional grids. In: Lynch, N.A., Shvartsman, A.A. (eds.) DISC 2010. LNCS, vol. 6343, pp. 297–311. Springer, Heidelberg (2010)
4. Cieliebak, M., Flocchini, P., Prencipe, G., Santoro, N.: Distributed computing by mobile robots: gathering. SIAM J. Comput. **41**, 829–879 (2012)
5. Cohen, R., Fraigniaud, P., Ilcinkas, D., Korman, A., Peleg, D.: Label-guided graph exploration by a finite automaton. ACM Trans. Algorithms **4**, 1–18 (2008)
6. Czyzowicz, J., Labourel, A., Pelc, A.: How to meet asynchronously (almost) everywhere. ACM Trans. Algorithms **8** (2012). article 37
7. Dereniowski, D., Pelc, A.: Drawing maps with advice. J. Parallel Distrib. Comput. **72**, 132–143 (2012)
8. Dessmark, A., Fraigniaud, P., Kowalski, D., Pelc, A.: Deterministic rendezvous in graphs. Algorithmica **46**, 69–96 (2006)
9. Dieudonné, Y., Pelc, A., Villain, V.: How to meet asynchronously at polynomial cost. In: Proceedings of the 32nd ACM Symposium on Principles of Distributed Computing (PODC 2013), pp. 92–99 (2013)
10. Ellis, R.: Volume of an N-simplex by multiple integration. Elem. Math. **31**, 57–59 (1976)
11. Emek, Y., Fraigniaud, P., Korman, A., Rosen, A.: Online computation with advice. Theor. Comput. Sci. **412**, 2642–2656 (2011)
12. Fraigniaud, P., Ilcinkas, D., Pelc, A.: Communication algorithms with advice. J. Comput. Syst. Sci. **76**, 222–232 (2010)
13. Fraigniaud, P., Ilcinkas, D., Pelc, A.: Tree exploration with advice. Inf. Comput. **206**, 1276–1287 (2008)
14. Fraigniaud, P., Korman, A., Lebhar, E.: Local MST computation with short advice. Theor. Comput. Syst. **47**, 920–933 (2010)

15. Fusco, E., Pelc, A.: Trade-offs between the size of advice and broadcasting time in trees. Algorithmica **60**, 719–734 (2011)
16. Gibbons, A.: Algorithmic Graph Theory. Cambridge University Press, Cambridge (1985)
17. Katz, M., Katz, N., Korman, A., Peleg, D.: Labeling schemes for flow and connectivity. SIAM J. Comput. **34**, 23–40 (2004)
18. Korman, A., Kutten, S., Peleg, D.: Proof labeling schemes. Distrib. Comput. **22**, 215–233 (2010)
19. Nisse, N., Soguet, D.: Graph searching with advice. Theor. Comput. Sci. **410**, 1307–1318 (2009)
20. Pelc, A.: Deterministic rendezvous in networks: a comprehensive survey. Networks **59**, 331–347 (2012)
21. Ta-Shma, A., Zwick, U.: Deterministic rendezvous, treasure hunts and strongly universal exploration sequences. In: Proceedings of the 18th ACM-SIAM Symposium on Discrete Algorithms (SODA 2007), pp. 599–608 2007
22. Thorup, M., Zwick, U.: Approximate distance oracles. J. ACM **52**, 1–24 (2005)

Computing the Dynamic Diameter
of Non-Deterministic Dynamic Networks
is Hard

Emmanuel Godard[✉] and Dorian Mazauric

Aix-Marseille Université, CNRS, LIF UMR 7279, 13000 Marseille, France
emmanuel.godard@lif.univ-mrs.fr

Abstract. A dynamic network is a communication network whose communication structure can evolve over time. The dynamic diameter is the counterpart of the classical static diameter, it is the maximum time needed for a node to causally influence any other node in the network. We consider the problem of computing the dynamic diameter of a given dynamic network. If the evolution is known a priori, that is if the network is deterministic, it is known it is quite easy to compute this dynamic diameter. If the evolution is not known a priori, that is if the network is non-deterministic, we show that the problem is hard to solve or approximate. In some cases, this hardness holds also when there is a static connected subgraph for the dynamic network.

In this note, we consider an important subfamily of non-deterministic dynamic networks: the time-homogeneous dynamic networks. We prove that it is hard to compute and approximate the value of the dynamic diameter for time-homogeneous dynamic networks.

1 Introduction

Highly Dynamic Networks. Most of existing research on networks and distributed computing has been devoted to *static* systems. The study of *dynamic* networks has focused extensively on systems where the dynamics are due to *faults* (e.g., node or edge deletions or additions); the faults however are limited in scope and bounded in number; and are considered anomalies with respect to the correct behaviour of the system. There are however systems where the instability never ends, the network is never connected, the changes are unbounded and occur continuously, where the changes are not anomalies but integral part of the nature of the system. Such *highly dynamic* systems are quite widespread, and becoming ubiquitous. The most common scenario is that of wireless mobile ad hoc networks, where the topology depends on the current distance between *mobile* nodes; typically, an edge exists at a given time if they are within communication range at that time. Hence, the topology changes continuously as the movement of the entities destroys old connections and creates new ones. These changes can be dramatic; connectivity does not necessarily hold, at least with the usual meaning of contemporaneous end-to-end multi-hop paths between

© Springer-Verlag Berlin Heidelberg 2015
J. Gao et al. (Eds.): ALGOSENSORS 2014, LNCS 8847, pp. 88–102, 2015.
DOI: 10.1007/978-3-662-46018-4_6

any pair of nodes, and the network may actually be disconnected at every time instant. These infrastructure-less highly dynamic networks, variously called *delay-tolerant, disruptive-tolerant, challenged, opportunistic*, have been long and extensively investigated by the engineering community and, more recently, by distributed computing researchers, especially with regards to the problems of broadcast and routing (*e.g.* [JLSW07, LW09, RR13, Zha06]). Interestingly, similar complex dynamics occur also in environments where there is no mobility at all, like in *social networks* (*e.g.* [CLP11, KKW08]).

Models and Dynamic Diameter. The highly dynamic features of these networks and their temporal nature is captured in a natural way by the (computationally equivalent) models of *evolving graphs* [Fer04] and of *time-varying graphs* [CFQS12], where edges between nodes exist only at some times. A crucial aspect of dynamic networks, and obviously of time-varying graphs, is that a path from a node to another might still exist over time, even though at no time the path exists in its entirety. It is this fact that renders routing, broadcasting, and thus computing possible in spite of the otherwise unsurmountable difficulties imposed by the nature of those networks. Hence, the notion of "path over time", formally called *journey* in [CFQS12], is a fundamental concept and plays a central role in the definition of almost all concepts related to connectivity in time-varying graphs. Examined extensively, under a variety of names (e.g., temporal path, schedule-conforming path, time-respecting path, trail), informally a journey is a walk $<e_1, e_2, ..., e_k>$ and a sequence of time instants $<t_1, t_2, ..., t_k>$ where edge e_i exists at time t_i.

The maximal length of such journeys is the counterpart of the diameter of classical static graphs. However, in dynamic networks, there are actually three such measures that one would like to minimize [CFQS12]:

- *foremost journey*: the journey that ends at the smallest round;
- *fastest journey*: the journey that minimizes the difference between the ending time and the starting time;
- *shortest journey*: the journey that minimizes the number of edges.

The maximal length of these journeys differs in most of the dynamic networks. In this note, we focus on the foremost journeys. The dynamic diameter is the maximum over all possible pairs of the length of the foremost journey. It corresponds to the maximum number of rounds needed for a node to causally influence any other node in the network. This value has also been called the *causal diameter*.

Given the erratic behaviour of a dynamic network, this value will, in general, depend on the moment in time. In this note, we consider a subfamily of dynamic networks that has the interesting property that the length of the foremost journey does not depend of the starting time. Informally, this family, the time-homogeneous dynamic networks, is defined so that the future possible behaviours are always the same during the evolution of the time. So this is a family well suited to investigate the dynamic diameter. More importantly, it should be noted that the results we obtain being mainly lower bounds, the hardness of approximation also applies to bigger families. We also argue that most interesting non-deterministic dynamic networks families must include one or more of the specific families considered here. See Sect. 1 for a more detailed discussion.

Families of Dynamic Networks. We consider discrete time evolution. The evolution of a dynamic network is described by a sequence of graphs. A non-deterministic dynamic network is defined by a set of such sequences. A dynamic network is time-homogeneous (or homogeneous) if its future evolution does not depend on what happened before. Such a dynamic network can be characterized by a set of graphs. Namely, there exists a set of graphs $\mathbf{G}$ such that at any moment, the instantaneous graph, that is the structure of the communication network that exists at a given moment, can be any graph $G \in \mathbf{G}$.

We will consider relevant families of homogeneous networks. We consider families of undirected (resp. directed) dynamic networks, i.e. $G \in \mathbf{G}$ is undirected (resp. directed). We will also consider statically connected (resp. strongly connected) dynamic networks, that is networks where there is a connected (resp. strongly connected) common spanning subgraph to any graph $G \in \mathbf{G}$.

Our results. We show that the computation of the dynamic diameter in non-deterministic dynamic networks is hard to compute, and in some cases it is hard to approximate. Even when there is a spanning subgraph that remains statically connected during the evolution of the network, the influence of the dynamic edges is difficult to evaluate. More precisely, we prove, by various reductions to the MAXIMUM INDUCED MATCHING problem, that:

- computing the dynamic diameter of undirected time-homogeneous dynamic networks is not in APX, *i.e.* it cannot be approximated within any constant factor in polynomial time, unless P = NP (Theorem 1);
- for statically connected time-homogeneous dynamic networks, the computation of the dynamic diameter is NP-complete (Theorem 2);
- when the network is directed and statically strongly connected, computing the dynamic diameter is not in APX (Theorem 3).

We were not able to prove or disprove that there exists an approximation algorithm for computing the dynamic diameter in undirected statically connected networks. The reductions we had to use suggest that it is not a simple matter in any direction.

The results are proved for time-homogeneous dynamic networks. In any family containing one of these families, it is also hard to compute the dynamic diameter. It should be noted that statically connected and strongly connected networks can be considered as the lesser dynamic in the families of dynamic networks. Any reasonnable dynamic networks family is expected to contain at least undirected statically connected networks. Moreover, the dynamicity of time-homogeneous networks being memoryless, that is a really basic dynamicity, any foreseeable non-deterministic dynamic networks family should admit time-homogeneous networks as a subfamily.

In other words, for non-sparse dynamic networks, it is highly unlikely that a family of interest would not contain one or more of the specific families considered here. This means that our results immediately apply to these families. For sparse dynamic networks (that is connected instantaneous graphs may not be admissible), depending of course on the specific model, it will probably be possible to extend our reductions by "decomposing" the graphs we use in sparser instantaneous graphs.

Note that since there are no common structure to describe arbitrary non-deterministic dynamic networks (the formalism of [CFQS12] does not specify how the presence function is encoded) from a complexity point of view, the time-homogeneous presentation is a one that really makes sense.

Related Work. In the case where the evolution is known a priori (the sequence of graphs is explicitly given) a polynomial algorithm is given in [BF03] for computing the dynamic diameter. In [BXFJ03], an algorithm is given for computing the fastest, shortest and foremost journeys. In [BF03], it is shown that computing the strongly connected components induced by a dynamic network is NP-complete. It was the only hardness result in this area until this year: in [MS14] the complexity of the Travelling Salesman Problem is investigated in the dynamic context; in [AKM14], the foremost coverage of recurrent time varying graphs is shown to be hard to approximate.

The statically connected networks family is the intersection of the $T-$interval connected families of dynamic networks introduced in [KLO10] to investigate Consensus. Given $T \in \mathbb{N}$, a $T-$interval connected network is such that for any interval of T rounds, there is a connected spanning subgraph common to the T instantaneous graphs.

Regarding the Consensus problem, following the work on benign faults in [CBS09], the message adversary model is introduced in [AG13]. This model actually corresponds to arbitrary dynamic networks where the communication primitive is a SENDALL operation (a node does not know to which of the other nodes the message is actually delivered). This does not affect the causal influence, therefore our results also directly apply to the message adversary model. In [RS13], it is proved that the message adversary model defines families that are computationally equivalent to lots of known and classical distributed models, including shared memory with failures. Time-homogeneous dynamic networks/message adversaries were considered in [GP11, CG13] where the equality between the time for broadcast and the time for solving Consensus is proved for some families. An immediate consequence of this note is that, given an arbitrary message adversary for which Consensus is solvable, the length (ie the time complexity) of the optimal solution is hard to approximate.

2   Notations and Definitions

Notations. In this note, we consider graphs, digraphs and dynamic graphs. Let $G = (V, E)$ be any graph with set of vertices V and set of edges E. We define $N_G(v) = \{u \mid v \in V, \{u, v\} \in E\}$ as the neighborhood of any node $v \in V$. We set $N_G[v] = N_G(v) \cup \{v\}$. We extend the notation to edges. Let $e = \{u, v\} \in E$. We define $N_G(e) = \{\{u, w\} \mid w \in V, \{u, w\} \in E\} \cup \{\{v, w\} \mid w \in V, \{v, w\} \in E\}$.

Let $D = (V, A)$ be any directed graph with set of vertices V and set of arcs A. We define $N_D^-(v) = \{u \mid (u, v) \in A\}$ as the in-neighborhood of any node $v \in V$. We set $N_D^-[v] = N_D^-(v) \cup \{v\}$. Similarly, we define $N_D^+(v) = \{u \mid (v, u) \in A\}$ as the out-neighborhood of any node $v \in V$. We set $N_D^+[v] = N_D^+(v) \cup \{v\}$.

A simple directed *path* c (resp. an undirected path $\bar{c}$) linking vertices u and v is a sequence of disjoint vertices $s_1, \ldots, s_k \in V$ where for all i, $1 \le i < k$, $(s_i, s_{i+1}) \in A$ (resp. $\{s_i, s_{i+1}\} \in E$), $s_1 = u$ and $s_k = v$. The *length of a path* c, denoted by $|c|$, is equal to the number of arcs composing it and the directed *distance* d (resp. undirected distance $\bar{d}$) between vertices is the length of the smallest simple directed (resp. undirected) path in G between u and v. A *strongly connected* digraph is a digraph D where the directed distance between any two vertices is finite.

Dynamic Communication Networks. We model an undirected (resp. directed) communication network by an undirected (resp. directed) graph $\mathcal{G} = (V, E)$ (resp. $\mathcal{G} = (V, A)$). We always assume that nodes have unique identities. Through this section, this graph $\mathcal{G}$ is fixed; it is called the *underlying graph*.

Communication in our model is reliable, and is performed in rounds, but with changing topology from round to round. Communication with a given topology is described by a spanning subgraph G of $\mathcal{G}$. We define the set $\Sigma = \{(V, E') \mid E' \subseteq E\}$ (resp. $\Sigma = \{(V, A') \mid A' \subseteq A\}$). This set represents all possible instantaneous communications given the underlying graph $\mathcal{G}$. For ease of notation, we will always identify a spanning subgraph in Σ with its set of edges (resp. arcs).

Definition 1. *An element G of Σ is called an* instantaneous graph. *A communication evolution (or* evolution*) is an infinite sequence $(G_i)_{i \in \mathbb{N}}$ of instantaneous graphs. A (non-deterministic)* dynamic network *is a set of communication evolutions.*

Given a set $\mathbf{G} \subseteq \Sigma$, we define $\mathcal{H}(\mathbf{G}) = \{(G_i)_{i \in \mathbb{N}} \mid G_i \in \mathbf{G}\}$ the set of all possible sequences of elements of $\mathbf{G}$.

Definition 2. *A dynamic network is* time-homogeneous *(or* homogeneous*) if there exists $\mathbf{G} \subseteq \Sigma$ such that the set of evolutions is $\mathcal{H}(\mathbf{G})$.*

Since we can describe a homogeneous dynamic network using $\mathcal{H}(\mathbf{G})$, for ease of notation, we will identify the dynamic network and $\mathbf{G}$. Given this set $\mathbf{G}$, it is possible to describe exactly all the possible evolutions for the corresponding time-homogeneous dynamic network.

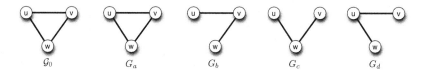

Fig. 1. Underlying graph $\mathcal{G}_0$, and its 4 possible instantaneous graphs.

Given an evolution $(G_i)_{i \in \mathbb{N}}$, the communication between two nodes u and v at a given round $i \in \mathbb{N}$ is possible if the the edge $\{u, v\} \in E(G_i)$ (resp. arc $(u, v) \in A(G_i)$). Broadcasting in such a system corresponds to "path over time", or journey.

Definition 3 (Journeys and Foremost Journeys). *Let $(G_i)_{i \in \mathbb{N}}$ be a communication evolution in $\mathbf{G}$. The sequence $(u_0, \cdots, u_p)$, $u_i \in V$, is a journey from u_0 to u_p with starting time i_0, and ending time $i_0 + p$ if for $0 \le k < p$, the edge $\{u_k, u_{k+1}\} \in E(G_{i_0+k})$ (resp. the arc $(u_k, u_{k+1}) \in A(G_{i_0+k})$) or $u_k = u_{k+1}$. Integer p is the length of the journey. A journey from u to v with starting time i_0 and length p is a foremost journey if any journey from u to v with starting time i_0 has length at least p.*

Definition 4. *Given $u, v \in V$, $d_{\mathbf{G}}^{i_0}$ denote the dynamic distance at time i_0 between u and v. It is the maximum length, in any evolution of $\mathbf{G}$, of the foremost journeys between u and v starting at time i_0.*

In homogeneous networks, the dynamic distance does not depend on the starting time, it is then denoted $d_{\mathbf{G}}(u, v)$. Note that, contrary to the general case of undirected dynamic networks, we have $d_{\mathbf{G}}(u, v) = d_{\mathbf{G}}(v, u)$ when $\mathbf{G}$ is homogeneous. As in the static case, the dynamic diameter is the maximum of dynamic distance over all possible pairs of vertices.

Definition 5 (Dynamic Diameter). $L(\mathbf{G}) = \max_{u,v \in V}(d_{\mathbf{G}}(u, v))$.

Equivalently, it is the maximum time needed to broadcast from any node of $\mathbf{G}$. In homogeneous networks, the diameter is defined if and only if all instantaneous graphs are connected. In the following, we always assume that the instantaneous graphs (resp. digraphs) are connected (resp. strongly connected).

Example of Homogeneous Dynamic Networks. On Fig. 1, we present the underlying graph $\mathcal{G}_0$ and its four connected instantaneous graphs G_a, G_b, G_c, G_d. Consider the homogeneous dynamic network $\mathbf{G}_0$ defined by $\{G_a, G_b, G_c\}$. An evolution of $\mathbf{G}_0$ is a sequence of graphs of $\{G_a, G_b, G_c\}$. The dynamic diameter of $\mathbf{G}_0$ is equal to 2.

We will compute the distances from u (by symmetry, we get the distances for v and w). If an evolution begins with G_a then v and w are reached. If an evolution begins with G_b, G_b then there is a journey (u, v) of length one to v, but you need a journey (u, v, w), of length 2, to get to w. Symmetrically, if the evolution begins with G_c, G_c, then the length is 2. If the evolution begins with G_b, G_c then you also need a journey, (u, u, w), of length 2, to get to w. So we get that $d_{\mathbf{G}_0}(u, v) = d_{\mathbf{G}_0}(u, w) = 2$ and the diameter $L(\mathbf{G}_0)$ of the dynamic network $\mathbf{G}_0$ is 2.

Static Connectivity.

Definition 6 (Statically Connected Networks). *Let* **G** *be an undirected dynamic network. An evolution* $(G_i)_{i \in \mathbb{N}}$ *is* statically connected *if* $\bigcap_{i \in \mathbb{N}} G_i$ *defines a connected graph. An undirected dynamic network* **G** *is* statically connected *if every evolution of* **G** *is statically connected.*

Similarly, we define statically strongly connected directed networks.

Definition 7 (Statically Strongly Connected Networks). *Let* **D** *be a directed dynamic network. An evolution* $(D_i)_{i \in \mathbb{N}}$ *is* statically strongly connected *if* $\bigcap_{i \in \mathbb{N}} D_i$ *defines a strongly connected digraph. A directed dynamic network* **D** *is* statically strongly connected *if every evolution of* **D** *is statically strongly connected.*

Note that statically connected networks are in general not time-homogeneous. But proving lower bounds for homogeneous statically connected networks extends obviously to all statically connected networks.

Maximum Induced Matching. We now define the MAXIMUM INDUCED MATCHING problem that is the problem that will be used in our reductions. A matching is a subset of edges that have no vertices in common. Intuitively, an induced matching corresponds to a subset of vertices such that its induced subgraph is a matching. This optimization problem has been proved to be NP-complete in [SV82]. It has been proved to be hard to approximate in [CC06].

Definition 8 (Maximum Induced Matching Problem). *Given a graph* $G = (V, E)$, *an* induced matching *is a set of edges* $E^* \subseteq E$ *such that* $|\{e' \in E^* \mid e' \in N_G(e)\}| \leq 1$ *for all* $e \in E$.

The MAXIMUM INDUCED MATCHING *problem consists in finding a set* $MIM(G)$ *that is an induced matching with maximal cardinality.*

3   The Dynamic Diameter Problem is Not in APX for Undirected Networks

In this section, we prove that the DYNAMIC DIAMETER problem is not in APX (Theorem 1). The reduction uses the MAXIMUM INDUCED MATCHING problem (Definition 8). Informally speaking, the reduction creates a correspondence between edges of the matching and instantaneous graphs such that when a graph corresponding to a given edge is used then all adjacent edges have a journey from

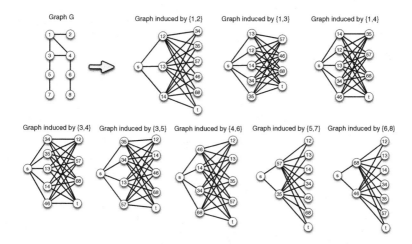

Fig. 2. Construction of the graphs $G_1, \ldots, G_8$ from a graph G with $|E| = 8$ edges.

a specific source s. Hence maximum foremost journeys must avoid creating adjacent edges.

Given any connected graph $G = (V, E)$ with at least 2 nodes, we construct the underlying graph $\mathcal{G} = (V', E')$ and the instantaneous graphs $G_1, \ldots, G_m$, where $m = |E|$. We set $\mathbf{G} = \{G_1, \ldots, G_m\}$ as the corresponding homogeneous dynamic network.

Let $E = \{e_1, e_2, \ldots, e_m\}$. Set $V' = \{s, t, u_1, \ldots, u_m\}$. Set $U_k = \{u_k\} \cup \{u_i \mid e_k \cap e_i \neq 0 \mid 1 \leq i \leq m\}$ and $E(G_k) = \cup_{u \in U_k} \{s, u\} \cup_{u \in U_k} \{u, t\} \cup_{u \in U_k, u' \in V' \setminus (U_k \cup \{s\})} \{u, u'\}$ for all k, $1 \leq k \leq m$. Set $E' = \cup_{1 \leq k \leq m} E(G_k)$. Figure 2 shows this construction from a simple undirected graph G.

Lemma 1. *Given any integer $K \geq 1$, if there exists an induced matching $IM(G)$ of G of size $|IM(G)| \geq K$, then $L(\mathbf{G}) \geq K + 1$.*

Proof. Suppose there exists an induced matching $IM(G)$ of G of size $|IM(G)| = K$. Without loss of generality, let $IM(G) = \{e_1, e_2, \ldots, e_K\}$ (we re-order the edges of G otherwise). Consider the evolution $S = (G_1, G_2, \ldots, G_K, G_1)$.

At the end of step 1, the set of nodes that have received a message is $N_{G_1}[s] = \{s\} \cup U_1$. At the end of step 2, the set of nodes that have received a message is $\{s\} \cup U_1 \cup U_2$ because $N_{G_2}(u) = U_2$ for all $u \in U_1$ and $N_{G_2}(s) = U_2$. Indeed, $U_1 \cap U_2 = \emptyset$ because $\{e_1, e_2, \ldots, e_K\}$ is an induced matching of G.

We now prove by induction that at the end of step k, $1 \leq k \leq K$, the set of nodes that have received a message is $\{s\} \cup U_1 \cup \ldots \cup U_k$. As previously shown, it is true for $k \in \{1, 2\}$. Suppose it is true for k, $1 \leq k \leq K - 1$. We prove that it is also true for $k + 1$. By induction hypothesis, the set of nodes that have received a message at the end of step k is $\{s\} \cup U_1 \cup \ldots \cup U_k$. Since $N_{G_{k+1}}(s) = U_{k+1}$ and $U_{k+1} \cap U_i = \emptyset$ for all i, $1 \leq i \leq k$, because $\{e_1, e_2, \ldots, e_K\}$ is an induced matching of G, then at the end of step $k + 1$, the set of nodes that have received a message is $\{s\} \cup U_1 \cup \ldots \cup U_{k+1}$. Thus, the result is true for $k + 1$.

In conclusion, at the end of step K, the set of nodes that have received a message is $\{s\} \cup U_1 \cup \ldots \cup U_K$. Since $t \notin U_i$ for all i, $1 \leq i \leq m$, then we need at least one more step to send a message to t. So $L(\mathbf{G}) \geq K + 1$.   □

Lemma 2. *Given any integer $K \geq 1$, if $L(\mathbf{G}) \geq K + 1$, then there exists an induced matching $IM(G)$ of G of size $|IM(G)| \geq K$.*

Proof. We prove that if any induced matching $IM(G)$ is of size $|IM(G)| \leq K - 1$, then $L(\mathbf{G}) \leq K$. Suppose that any induced matching $IM(G)$ is of size $|IM(G)| \leq K - 1$. Consider any sequence $S = (G_{i_1}, \ldots, G_{i_K})$ of graphs with $1 \leq i_j \leq m$ for all j, $1 \leq j \leq K$. By hypothesis, there exists a node $u \in V'$ such that $u \in U_{i_j}$ and $u \in U_{i_{j'}}$ for some j, j', $1 \leq j < j' \leq K$. Thus, at step j, u receives a message. Furthermore, at step j', u sends a message to all nodes of $V' \setminus U_{i_{j'}}$, and s sends a message to all nodes of $U_{i_{j'}}$ because $N_{G_{i_{j'}}}(s) = U_{i_{j'}}$. In conclusion, all nodes have received a message at step $j' \leq K$, and so $L(\mathbf{G}) \leq K$.   □

We are now able to prove the main result of this section.

Theorem 1. *The* Dynamic Diameter *problem on undirected time-homogeneous dynamic networks is not in APX.*

Proof. Lemmas 1 and 2 prove that for any graph G it is possible to construct a dynamic network $\mathbf{G}$ of size that is polynomial in $|G|$, such that $L(\mathbf{G}) = |MIM(G)| + 1$ where $MIM(G)$ is a maximum cardinality induced matching of G. Since the Maximum Induced Matching problem is not in APX [CC06], then the Dynamic Diameter problem for undirected networks is not in APX.   □

4   The Dynamic Diameter Problem is NP-Complete Even for Statically Connected Networks

This section is devoted to proving that the Dynamic Diameter problem is NP-complete even if the undirected dynamic network is statically connected (Theorem 2). We use in our reduction the Maximum Induced Matching problem (Definition 8). Informally speaking, the reduction creates again a correspondence between edges of the matching and instantaneous graphs such that when a graph corresponding to a given edge is used then all adjacent edges have a journey from a specific source s. Therefore maximum foremost journeys must avoid creating adjacent edges. It has to be noted that the reduction does depend on the size K of the Maximum Induced Matching, therefore we can only obtain NP-completeness from this reduction.

Given any connected graph $G = (V, E)$ with at least 2 nodes, and $K \geq 2$ be any integer, we construct the underlying graphs $\mathcal{G}^K = (V', E')$ and the instantaneous graphs $G_1, \ldots, G_m$ where $m = |E|$. $\mathbf{G}^K = \{G_1, \ldots, G_m\}$ is the corresponding homogeneous dynamic network.

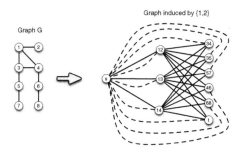

Fig. 3. Construction of the graph G_1 from a graph G with $|E| = 8$ edges. The curved edges represent the length K paths $P_1, \ldots, P_9$.

Let $E = \{e_1, e_2, \ldots, e_m\}$. Let $P_1, \ldots, P_{m+1}$ be $m+1$ disjoint paths composed of $K + 1$ nodes. Let $s_1, \ldots, s_{m+1}$ be the first nodes of the paths $P_1, \ldots, P_{m+1}$, respectively. Let $t_1, \ldots, t_{m+1}$ be the last nodes of the paths $P_1, \ldots, P_{m+1}$, respectively.

Let $V_i(P_k)$ be the set of the first i nodes (including s_k) of path P_k for all i, $1 \leq i \leq K + 1$, and for all k, $1 \leq k \leq m + 1$. Let $V_i'(P_k)$ be the set of the last i nodes (including t_k) of path P_k for all i, $1 \leq i \leq K + 1$, and for all k, $1 \leq k \leq m + 1$.

Set $V' = \{s, t, u_1, \ldots, u_m\} \cup V(P_1) \cup \ldots \cup V(P_{m+1})$. Set $U_k = \{u_k\} \cup \{u_i \mid e_k \cap e_i \neq \emptyset \mid 1 \leq i \leq m\}$ and set $U = \{u_1, \ldots, u_m\}$. Set $E(G_k) = \cup_{u \in U_k} \{s, u\} \cup_{u \in U_k, u' \in U \setminus U_k} \{u, u'\} \cup_{1 \leq i \leq m+1} E(P_i) \cup_{1 \leq i \leq m+1} \{s, s_i\} \cup_{1 \leq i \leq m} \{t_i, u_i\} \cup \{t_{m+1}, t\}$ for all k, $1 \leq k \leq m$. Set $E' = \cup_{1 \leq k \leq m} E(G_k)$.

Figure 3 shows the construction of G_1 from an undirected graph G.

Lemma 3. *If there exists an induced matching $IM(G)$ of G of size $|IM(G)| \geq K - 1$, then $L(\mathbf{G}^K) \geq K + 1$.*

Proof. Suppose there exists an induced matching $IM(G)$ of G of size $|IM(G)| = K - 1$. Without loss of generality, let $IM(G) = \{e_1, e_2, \ldots, e_{K-1}\}$ (we re-order the edges of G otherwise).

Consider the evolution $S = (G_1, G_2, \ldots, G_{K-1}, G_1, G_1)$. At the end of step 1, the set of nodes that have received a message is $N_{G_1}[s] = \{s, s_1, \ldots, s_{m+1}\} \cup U_1$. At the end of step 2, the set of nodes that have received a message is $\{s\} \cup V_2(P_1) \cup \ldots \cup V_2(P_{m+1}) \cup_{j \mid u_j \in U_1} t_j \cup U_1 \cup U_2$ because $N_{G_2}(u_j) = U_2 \cup t_j$ for all $u_j \in U_1$ and $N_{G_2}(s) = U_2 \cup \{s_1, \ldots, s_{m+1}\}$. Indeed, $U_1 \cap U_2 = \emptyset$ because $\{e_1, e_2, \ldots, e_K\}$ is an induced matching of G.

We now prove by induction that at the end of step k, $1 \leq k \leq K - 1$, the set of nodes that have received a message is $\{s\} \cup V_k(P_1) \cup \ldots \cup V_k(P_{m+1}) \cup_{j \mid u_j \in U_1} V_{k-1}'(P_j) \cup_{j \mid u_j \in U_2} V_{k-2}'(P_j) \ldots \cup_{j \mid u_j \in U_{k-1}} V_1'(P_j) \cup U_1 \cup \ldots \cup U_k$. Consider that $U_j = \emptyset$ if $j \notin \{1, 2, \ldots, m\}$ and that $V_j'(P) = \emptyset$ for any path $P \in \{P_1, \ldots, P_{m+1}\}$ if $j < 1$. As previously shown, it is true for $k \in \{1, 2\}$. Suppose it is true for k, $1 \leq k \leq K - 2$. We prove that it is also true for $k+1$. By induction hypothesis, the set of nodes that have received a message at the end of step k is $\{s\} \cup V_k(P_1) \cup \ldots \cup$

$V_k(P_{m+1})\cup_{j|u_j\in U_1}V'_{k-1}(P_j)\cup_{j|u_j\in U_2}V'_{k-2}(P_j)\ldots\cup_{j|u_j\in U_{k-1}}V'_1(P_j)\cup U_1\cup\ldots\cup U_k$.
Since $N_{G_{k+1}}(s)=\{s_1,\ldots,s_{m+1}\}\cup U_{k+1}$ and $U_{k+1}\cap U_i=\emptyset$ for all i, $1\le i\le k$
because $\{e_1,e_2,\ldots,e_K\}$ is an induced matching of G, then at the end of step $k+1$,
the set of nodes that have received a message is $\{s\}\cup U_1\cup\ldots\cup U_{k+1}$ union the set
of nodes that have received a message that belong to $V(P_1)\cup\ldots\cup V(P_{m+1})$. This
set corresponds to $\cup_{j|u_j\in U_1}V'_k(P_j)\cup_{j|u_j\in U_2}V'_{k-1}(P_j)\ldots\cup_{j|u_j\in U_k}V'_1(P_j)$. Thus,
the result is true for $k+1$.

In conclusion, at the end of step $K-1$, the set of nodes that have received
a message is $\{s\}\cup V_{K-1}(P_1)\cup\ldots\cup V_{K-1}(P_{m+1})\cup_{j|u_j\in U_1}V'_{K-2}(P_j)\cup_{j|u_j\in U_2}$
$V'_{K-3}(P_j)\ldots\cup_{j|u_j\in U_{K-2}}V'_1(P_j)\cup U_1\cup\ldots\cup U_{K-1}$. Node t does not belong to this
set by construction of the sequence and because each path P_i has $K+1$ nodes,
for all i, $1\le i\le m+1$. Thus, the set of nodes $V'_2(P_{m+1})$ have not received a
message. Therefore we need one more step to send a message to node t and to
the node of $V'_2(P_{m+1})\setminus\{t_{m+1}\}$; and we need one another more step to send a
message to node t_{m+1}. Thus, $L(\mathbf{G}^K)\ge K+1$.   □

Lemma 4. *If $L(\mathbf{G}^K)\ge K+1$, then there exists an induced matching $IM(G)$
of G of size $|IM(G)|\ge K-1$.*

Proof. We prove that if any induced matching $IM(G)$ is of size $|IM(G)|\le
K-2$, then $L(\mathbf{G}^K)\le K$. Suppose that any induced matching $IM(G)$ is of
size $|IM(G)|\le K-2$. Consider any sequence $S=(G_{i_1},\ldots,G_{i_{K-1}})$ of graphs
with $1\le i_j\le m$ for all j, $1\le j\le K-1$. By hypothesis, there exists a node
$u\in V'$ such that $u\in U_{i_j}$ and $u\in U_{i_{j'}}$ for some j,j', $1\le j<j'\le K-1$.
Thus, at step j, u receives a message. Furthermore, at step j', u sends a message
to all nodes of $\{t\}\cup(U\setminus U_{i_{j'}})$, and $\{s\}$ sends a message to all nodes of $U_{i_{j'}}$
because $U_{i_{j'}}\subset N_{G_{i_{j'}}}(s)$. This implies that at the end of step K, the set of
nodes $\{t_1,\ldots,t_{m+1}\}$ have received a message. Furthermore, the set of nodes
$V_K(P_1)\cup\ldots\cup V_K(P_{m+1})$ have received a message at the end of step K because
the paths have all size K.

In conclusion, all nodes have received a message at the end of step K, and
so $L(\mathbf{G}^K)\le K$.   □

Theorem 2. *Solving the* DYNAMIC DIAMETER *problem on the set of statically
connected time-homogeneous dynamic networks is NP-complete.*

Proof. Noting that the paths $P_1,\cdots,P_{m+1}$ are static, it is immediate that $\mathbf{G}^K$
is statically connected. Lemmas 3 and 4 prove that for any graph G it is possible
to construct a dynamic network $\mathbf{G}^K$ of size that is polynomial in $|G|$, such
that $L(\mathbf{G}^K)\ge K+1$ if, and only if, $|MIM(G)|\ge K-1$, where $MIM(G)$
is a maximum cardinality induced matching of G. More over $\mathbf{G}^K$ is statically
connected.

From [BF03], checking the maximum of the dynamic distance of a given
evolution can be done in polynomial time, therefore the DYNAMIC DIAMETER
problem is in NP. Since the MAXIMUM INDUCED MATCHING problem is NP-
complete [SV82], then the DYNAMIC DIAMETER problem is NP-complete on the
family of statically connected networks.   □

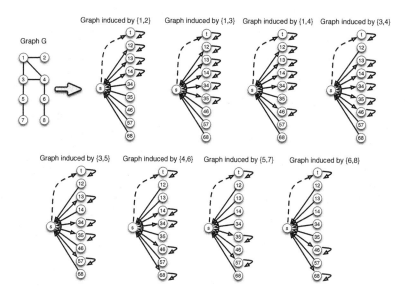

Fig. 4. The digraphs D and $D_1, \ldots, D_m$ constructed from the graph G. The dotted arcs represent the path from s to t and the curved arcs join all (other) nodes.

5 The Dynamic Diameter Problem is Not in APX for Statically Strongly Connected Networks

In this section, we prove that the DYNAMIC DIAMETER problem is not in APX even for statically connected directed networks (Theorem 3). The reduction uses the MAXIMUM INDUCED MATCHING problem (Definition 8). The reduction here is a bit different, since the correspondence between instantaneous graphs and edges of the matching is much more direct. Note that, we do need the directed links in order to maintain very closely this correspondence.

Given any connected graph $G = (H, E)$ with at least 2 nodes, we construct the underlying digraph $\mathcal{D} = (V, A)$ and the instantaneous digraphs $D_1, \ldots, D_m$ where $m = |E|$. $\mathbf{D} = \{D_1, \ldots, D_m\}$ is the corresponding homogeneous dynamic network.

Let $E = \{e_1, e_2, \ldots, e_m\}$. Set $V = \{s, v_1, v_2, \ldots v_m, t, u_1, u_2, \ldots u_m\}$. Let $U_i = \{u_j \mid u_i \cap u_j \neq \emptyset \mid 1 \leq j \leq m\}$ for all i, $1 \leq i \leq m$, and let $U = U_1 \cup \ldots \cup U_m$. Set $A(D_i) = \{(s, v_1), (v_m, t)\} \cup \{(v_j, v_{j+1}) \mid 1 \leq j \leq m - 1\} \cup \{(t, u) \mid u \in V \setminus \{t\}\} \cup \{(u_j, s) \mid 1 \leq j \leq m\} \cup \{(s, u) \mid u \in U_i\} \cup \{(u, u') \mid u \in U_i, u' \in V \setminus U_i\}$ for all i, $1 \leq i \leq m$. Set $A = \cup_{1 \leq i \leq m} A(D_i)$. Figure 4 shows this construction from a simple undirected graph G. Note that for all i, $1 \leq i \leq m$, D_i is strongly connected. Furthermore, the graph induced by V and the set of arcs $\{(s, v_1), (v_1, v_2), \ldots, (v_{m-1}, v_m), (v_m, t)\} \cup \{(t, u) \mid u \in V \setminus \{t\}\} \cup \{(u_j, s) \mid 1 \leq j \leq m\}$ is strongly connected and appears in all D_i's.

Lemma 5. *Given any induced matching $IM(G)$ of G, then $L(\mathbf{D}) \geq |IM(G)| + 1$.*

Proof. Let $IM(G)$ be any induced matching of G of size $K \geq 1$. Without loss of generality, let $IM(G) = \{e_1, e_2, \ldots, e_K\}$ (we re-order the nodes of $\mathcal{D}$ otherwise). Consider the evolution $S = (D_1, D_2, \ldots, D_K)$.

At the end of step 1, the set of nodes that have received a message is $N_{D_1}^+[s] = \{s, v_1\} \cup U_1$. At the end of step 2, the set of nodes that have received a message is $\{s, v_1, v_2\} \cup U_1 \cup U_2$ because $N_{D_2}^+(u) = \{s\}$ for all $u \in U_1$ and because $N_{D_2}^+(v_1) = \{v_2\}$. Indeed, $U_1 \cap U_2 = \emptyset$ because $\{e_1, e_2, \ldots, e_K\}$ is an induced matching of G.

We now prove by induction that at the end of step k, $1 \leq k \leq K$, the set of nodes that have received a message is $\{s, v_1, \ldots, v_k\} \cup U_1 \cup \ldots \cup U_k$. As previously shown, it is true for $k \in \{1, 2\}$. Suppose it is true for k, $1 \leq k \leq K - 1$. We prove that it is also true for $k + 1$. By induction hypothesis, the set of nodes that have received a message at the end of step k is $\{s, v_1, \ldots, v_k\} \cup U_1 \cup \ldots \cup U_k$. Since $\{e_1, e_2, \ldots, e_K\}$ is an induced matching of G, then $N_{D_{k+1}}^+(u) = \{s\}$ for all $u \in U_1 \cup \ldots \cup U_k$ because $U_{k+1} \cap U_i = \emptyset$ for all i, $1 \leq i \leq k$. Furthermore, $N_{D_{k+1}}^+(v_i) = v_{i+1}$ for all i, $1 \leq i \leq k \leq K - 1 \leq m - 1$. Finally, since $N_{D_{k+1}}^+(s) = \{v_1\} \cup U_{k+1}$, then at the end of step $k + 1$, the set of nodes that have received a message is $\{s, v_1, \ldots, v_{k+1}\} \cup U_1 \cup \ldots \cup U_{k+1}$.

In conclusion, at the end of step K, the set of nodes that have received a message is $\{s, v_1, \ldots, v_K\} \cup U_1 \cup \ldots \cup U_K$. Since $t \notin U_i$ for all i, $1 \leq i \leq m$ and $t \neq v_j$ for all j, $1 \leq j \leq m$, then we need at least one more step to send a message to t. Thus, $L(\mathbf{D}) \geq K + 1 = |IM(G)| + 1$. □

Lemma 6. *There exists an induced matching* $IM(G)$ *of* G *of size* $|IM(G)| \geq L(\mathbf{D}) - 1$.

Proof. We prove the result by contradiction. Suppose that $|IM(G)| \leq L(\mathbf{D}) - 2$ for any induced matching $IM(G)$ of G. Thus, in any sequence $D_{i_1}, D_{i_2}, \ldots, D_{i_{L(D)-1}}$, then there exist i_j and $i_{j'}$ $(1 \leq i_j, i_{j'} \leq m)$, $1 \leq j < j' \leq L(\mathbf{D}) - 1$, such that $U_{i_j} \cap U_{i_{j'}} \neq \emptyset$ by construction of the digraphs D_i for all i, $1 \leq i \leq m$. Thus, at the end of step j', all nodes of V have received a message. A contradiction. In conclusion, there exists an induced matching $IM(G)$ of G of size $|IM(G)| \geq L(\mathbf{D}) - 1$. □

Theorem 3. *The* DYNAMIC DIAMETER *problem on the set of statically strongly connected networks is not in APX.*

Proof. Lemmas 5 and 6 prove that for any graph G it is possible to construct a dynamic network $\mathbf{D}$ of size that is polynomial in $|G|$, such that $L(\mathbf{D}) = |MIM(G)| + 1$ where $MIM(G)$ is a maximum cardinality induced matching of G. Furthermore, $\mathbf{D}$ is statically strongly connected. Since the MAXIMUM INDUCED MATCHING problem is not in APX [CC06], then the DYNAMIC DIAMETER problem is not in APX even for statically strongly connected networks. □

6 Conclusion and Future Works

In this note, we proved that computing the dynamic diameter of homogeneous dynamic networks is hard, even when there exists a connected (or strongly connected) spanning subgraph that is static. In the directed case, we were able to prove that the problem has no constant factor approximation polynomial algorithm, unless P = NP. Note that according to recent almost tight results on the maximum induced matching [CLN13], we obtain by our reduction a similar tightness on the (non-)approximability of the dynamic diameter problem.

We were not able to prove the non-approximability in the case of undirected dynamic networks with static connectivity. Whether there exists an approximation algorithm is an open question in this case. We observe that proving that computing the dynamic diameter in undirected statically connected networks is not in APX would imply Theorems 1 and 3.

Another interesting question would be to consider the problem of finding the maximal length of the fastest and shortest journeys. Note that in the setting of time-homogeneous dynamic networks, as in the general setting, the length of foremost and fastest journeys can differ.

References

[AG13] Afek, Y., Gafni, E.: Asynchrony from synchrony. In: Frey, D., Raynal, M., Sarkar, S., Shyamasundar, R.K., Sinha, P. (eds.) ICDCN 2013. LNCS, vol. 7730, pp. 225–239. Springer, Heidelberg (2013)

[AKM14] Aaron, E., Krizanc, D., Meyerson, E.: DMVP: foremost waypoint coverage of time-varying graphs. In: Kratsch, D., Todinca, I. (eds.) WG 2014. LNCS, vol. 8747, pp. 29–41. Springer, Heidelberg (2014)

[BF03] Bhadra, S., Ferreira, A.: Complexity of connected components in evolving graphs and the computation of multicast trees in dynamic networks. In: Pierre, S., Barbeau, M., An, H.-C. (eds.) ADHOC-NOW 2003. LNCS, vol. 2865, pp. 259–270. Springer, Heidelberg (2003)

[BXFJ03] Bui-Xuan, B.-M., Ferreira, A., Jarry, A.: Computing shortest, fastest, and foremost journeys in dynamic networks. Int. J. Found. Comput. Sci. 14(2), 267–285 (2003)

[CBS09] Charron-Bost, B., Schiper, A.: The heard-of model: computing in distributed systems with benign faults. Distrib. Comput. 22(1), 49–71 (2009)

[CC06] Chlebík, M., Chlebíková, J.: Complexity of approximating bounded variants of optimization problems. Theor. Comput. Sci. 354(3), 320–338 (2006)

[CFQS12] Casteigts, A., Flocchini, P., Quattrociocchi, W., Santoro, N.: Time-varying graphs and dynamic networks. Int. J. Parallel, Emerg. Distrib. Syst. 27(5), 387–408 (2012)

[CG13] Coulouma, É., Godard, E.: A characterization of dynamic networks where consensus is solvable. In: Moscibroda, T., Rescigno, A.A. (eds.) SIROCCO 2013. LNCS, vol. 8179, pp. 24–35. Springer, Heidelberg (2013)

[CLN13] Chalermsook, P., Laekhanukit, B., Nanongkai, D.: Independent set, induced matching, and pricing: connections and tight (subexponential time) approximation hardnesses. In: FOCS, pp. 370–379. IEEE Computer Society (2013)

[CLP11] Chierichetti, F., Lattanzi, S., Panconesi, A.: Rumor spreading in social networks. Theor. Comput. Sci. 412(24), 2602–2610 (2011)

[Fer04] Ferreira, A.: Building a reference combinatorial model for MANETs. IEEE Netw. 18(5), 24–29 (2004)

[GP11]   Godard, E., Peters, J.: Consensus vs. broadcast in communication networks with arbitrary mobile omission faults. In: Kosowski, A., Yamashita, M. (eds.) SIROCCO 2011. LNCS, vol. 6796, pp. 29–41. Springer, Heidelberg (2011)

[JLSW07] Jones, E.P.C., Li, L., Schmidtke, J.K., Ward, P.A.S.: Practical routing in delay-tolerant networks. IEEE Trans. Mob. Comput. 6(8), 943–959 (2007)

[KKW08] Kossinets, G., Kleinberg, J., Watts, D.: The structure of information pathways in a social communication network. In: Proceedings of the 14th International Conference on Knowledge Discovery and Data Mining (KDD), pp. 435–443 (2008)

[KLO10] Kuhn, F., Lynch, N., Oshman, R.: Distributed computation in dynamic networks. In: Proceedings of the 42nd ACM Symposium on Theory of computing (STOC), pp. 513–522. ACM (2010)

[LW09]   Liu, C., Wu, J.: Scalable routing in cyclic mobile networks. IEEE Trans. Parallel Distrib. Syst. 20(9), 1325–1338 (2009)

[MS14]   Michail, O., Spirakis, P.G.: Traveling salesman problems in temporal graphs. In: Csuhaj-Varjú, E., Dietzfelbinger, M., Ésik, Z. (eds.) MFCS 2014, Part II. LNCS, vol. 8635, pp. 553–564. Springer, Heidelberg (2014)

[RR13]   Ros, F.J., Ruiz, P.M.: Minimum broadcasting structure for optimal data dissemination in vehicular networks. IEEE Trans. Veh. Technol. 62(8), 3964–3973 (2013)

[RS13]   Raynal, M., Stainer, J.: Synchrony weakened by message adversaries vs asynchrony restricted by failure detectors. In: Fatourou, P., Taubenfeld, G. (eds.) PODC, pp. 166–175. ACM (2013)

[SV82]   Stockmeyer, L., Vazirani, V.: NP-completeness of some generalizations of the maximum matching problem. Inf. Process. Lett. 15(1), 14–19 (1982)

[Zha06] Zhang, Z.: Routing in intermittently connected mobile ad hoc networks and delay tolerant networks: overview and challenges. IEEE Commun. Surv. Tutor. 8(1), 24–37 (2006)

Improved Spanners in Networks
with Symmetric Directional Antennas

Stefan Dobrev[1][(⊠)] and Milan Plžík[2]

[1] Department of Informatics, Institute of Mathematics,
Slovak Academy of Sciences, Bratislava, Slovakia
`stefan.dobrev@savba.sk`
[2] Department of Computer Science, Comenius University, Bratislava, Slovakia

Abstract. Consider a set S of sensors in a plane, each equipped with a directional antenna of beamwidth $\frac{\pi}{2}$ and radius $r \in O(1)$. The antennas are symmetric, i.e. for two sensors u and v to communicate, u has to be covered by the antenna of v and vice versa. Assuming the Unit Disc Graph (UDG) of S is connected, the problem we attack is: How to orient the antennas so that the resulting connectivity graph is a k-spanner of the UDG, while minimizing the radius used?

We provide two results: (a) 7-spanner using radius 33, and (b) 5-spanner, still using $O(1)$ radius. This significantly improves the previous state of the art (8-spanner), even improving upon the pre previous best result (6-spanner) for beamwidth $\frac{2\pi}{3}$.

Keywords: Sensor networks · Directional antenna · Graph spanners · Stretch factor

1 Introduction

Wireless networks have traditionally employed omnidirectional antennas for transmission and communication. These antennas can be relatively simple to model using disks centered at the transmitter locations, with the disk radius corresponding the power of transmission. However, omnidirectional antennas do have drawbacks, most notably potential waste of energy as the transmission is not necessarily focused where it is needed most; omnidirectional transmission also leads to greater interference and signal degradation. This has motivated engineering research into directional antennas, resulting in great advances towards practical use of directional antennas in wireless networks [13] in recent years.

The coverage area of a directional antenna is specified by its *beamwidth* φ, *range* r, location l and direction $\boldsymbol{d}$: The set of covered point is a cone with the tip at l, with the central axis pointing at direction $\boldsymbol{d}$, of width φ and range r. The locations are fixed in static networks (e.g. represented by a point set S in the Euclidean plane), the beamwidth is a constant given by the technology/hardware, range is determined by the transmission power and the typical

Supported by VEGA 2\0136\12.

J. Gao et al. (Eds.): ALGOSENSORS 2014, LNCS 8847, pp. 103–121, 2015.
DOI: 10.1007/978-3-662-46018-4_7

goal is to minimize it as much as possible. This leaves the orientation as the main degree of freedom for the network designer.

Historically, two models of communication with directional antennas have been studies: *asymmetric* and *symmetric* one. In the better studied model of asymmetric communication, the transmitters are directional, while the receivers are omnidirectional. This leads to a situation, where communication links can be directional, with u being able to transmit to v, but not vice versa, as the v's transmitter can point away from u. There has been considerable research interest in this area, investigating the tradeoffs between the number of antennas per node, their beamwidth and transmission range, while trying to ensure strong connectivity of the resulting directed communication graph – see [10] for a survey.

In the model of symmetric communication (using so-called DD-antennas - Directional transmission, Directional receiving), u and v can communicate if an only if u lies in the coverage area of v's antenna, and v lies in the coverage area of u's antenna. The resulting communication graph is therefore an undirected graph (Fig. 1).

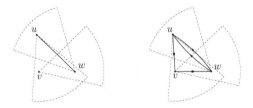

Fig. 1. Symmetric model (left) and asymmetric model (right). Solid lines represent communication links.

In this paper we consider the symmetric communication model. The antenna locations are represented by a set S of points in the 2D Euclidean plane. Each point (sensor) is equipped with a directional antenna of beamwidth $\frac{\pi}{2}$ and range $r \in O(1)$, i.e. the beamwidth and ranges of all antennas are the same. The antennas differ in their orientation – in fact, our main goal is to orient each antenna so that the desired network property is achieved.

We assume that the points set S is such that the unit disc graph $\mathrm{UDG}(S)$ is connected. Our goal is to find an antenna assignment A specifying for each antenna its orientation, such that the communication graph $\mathrm{IG}(S, A)$ induced on S by A is a k-spanner of $\mathrm{UDG}(S)$ (i.e. for each edge $(u, v) \in \mathrm{UDG}(S)$, there is a path of length at most k in $\mathrm{IG}(S, A)$). Our aim is to find the smallest possible k, while keeping the radius r as low as possible.

1.1 Related Work

The first results for directional antennas were achieved in the asymmetric communication model. The authors in [4] investigated the strong connectivity problem when each sensor is equipped with a single directional antenna. They showed that the problem is NP-hard for beamwidth less than $\frac{2\pi}{3}$ and provided several algorithms to approximate the minimum transmission range for larger beam-width. Subsequent research investigated the strong connectivity problem in slightly different settings, considering several antennas per node [8], as well as double or dipole antennas [9]. The problem of finding a k-spanner, also called the stretch factor problem, has received considerable attention in the asymmetric model [3,6,11] as well. A comprehensive survey of the connectivity problem for directional antennas can be found in [10].

The results for the symmetric directional antennas are more recent and less exhaustive. In the first paper investing symmetric directional antenna [5], the authors show that beamwidth $\varphi \geq \pi/3$ is sufficient to guarantee connectivity, however, the radius used is related to the diameter of the whole point set. They prove that sometimes $\varphi = \pi/3$ is necessary, but do not investigate the complexity of determining whether there exists an antenna orientation ensuring connectivity, while satisfying given restrictions on φ and antenna range. That problem has been addressed in [1] and [7], where it has been shown to be NP-hard for some specific combinations of φ and r. The stretch factor problem has also been addressed there, providing a range of results: a 9-spanner construction using radius 10 for $\varphi = \frac{\pi}{2}$, a 6-spanner using radius 7 for $\varphi = \frac{2\pi}{3}$ (this has been shown independently in [1] as well) and a 3-spanner using radius 5 for $\varphi = \pi$. [2] contains several related results, among them the most relevant to us is a construction of an 8-spanner using radius $14\sqrt{2}$.

1.2 Our Results

In this paper, we focus on improving the results of [2,7] for the stretch factor problem in the case of single $\varphi = \frac{\pi}{2}$. We combine previous approaches with novel ideas and techniques to significantly improve upon the state of the art for the stretch factor problem in this setting. In fact, the stretch factor of 5 that we are able to achieve for $\varphi = \frac{\pi}{2}$ beats the best stretch factor achieved so far for beamwidth of $\frac{2\pi}{3}$. We pay in the radius and the complexity of the construction, though.

Our results in the context of previous work are captured in Table 1:

Table 1. Results for the stretch-factor problem for antennas of beamwidth $\pi/2$.

Radius	Stretch Factor	Source
10	9	[7]
$14\sqrt{2}$	8	[2]
33	7	This work
$O(1)$	5	This work

1.3 Preliminaries and Notation

We consider a point-set S (called sensors or vertices) in two-dimensional Euclidean space (whose elements, regardless of whether occupied by sensors, we will call points). We assume the unit disc graph of S (UDG(S)) is connected and measure the antenna radius in terms of the longest edge of its MST.

Throughout this paper we use the following notation:

- $CE(P)$ is a convex envelope of a point-set P,
- $d_e(P_1, P_2)$ is the shortest Euclidean distance between point-sets P_1 and P_2;
- $d_h(P_1, P_2)$ is the shortest hop distance between P_1 and P_2 in UDG(S),
- $D_e(P_1, P_2) = \max_{u \in P_1, v \in P_2} d_e(u, v)$, D_h is defined analogously using d_h.
- $D_e(P)$ is the Euclidean diameter of P, $D_h(P)$ is the hop-distance diameter of P.

Let $v \in S$. We will use $r(v)$, $l(v)$ and $c(v)$ to denote the counter-clockwise, clockwise and central rays of v's antenna, respectively. We will use "u sees v" as a shorthand for "the antenna of u covers v".

Let P be a vertex set. A function $A : P \mapsto \langle 0, 2\pi \rangle$ assigning to each vertex of the set P the direction of the central ray of its antenna will be called an *antenna assignment* on P.

2   5-Gadgets and their Properties

2.1   Outline

The approach taken by [2] and [7] can be summarized as follows:

(i) show that it is possible to orient the antennas of four sensors in such a way, that they cover the whole plane

(ii) show that if two such quadruples are separated by a line, then there is a pair of sensors, each from a different quadruple, that sees it other

(iii) construct a maximal set of quadruples such that each quadruple has small geometric radius and any two quadruples are separated by a line

(iv) use the quadruples as hubs, orient non-hub sensors to nearby hubs and use (ii) to connect nearby hubs.

Since the hubs have hop diameter 3, an edge (u, v) with u and v connecting to different hubs has stretch 9 (1 to connect u to its hub, 3 within the hub, 1 to connect to the other hub, 3 within that hub, and 1 to reach v). This can be slightly reduced to 8 in [2] by careful selection of the rules how to select quadruples and connecting non-hub sensors to them, at the expense of needing a larger radius compared to [7].

The basic idea of our approach is to follow the same strategy, but instead of using 4-sensor hubs (which inevitably must have diameter 3 in the worst case) to employ 5-sensor hubs. We will show these can be constructed with hop diameter 2, allowing us to reduce the stretch factor to 7. Further reduction to stretch 5 employs a new technique to deal with the problematic (having stretch 7 using the above mentioned approach) edges.

2.2   5-Gadgets

There are several technical obstacles on our path, the first one being that we can make our construction work only if the hub points are in convex position. To ensure that, we employ the following result:

Theorem 1. ([12]) *Any set of 9 non-collinear points in a plane contains 5 points forming a convex polygon.*

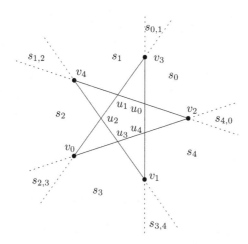

Fig. 2. Simple 5-gadget, with marked main and side sectors

Let $P = \{v_0, v_1, v_2, v_3, v_4\}$ be a set of 5 vertices in convex position, oriented counter-clockwise – we will call it a *5-gadget*. We use $i \oplus j$ to denote $(i + j)$ mod 5 and similarly, $i \ominus j$ to denote $(i - j)$ mod 5. Let $v_i \in P$. We say $v_{i \oplus 1}$ is a *successor* of v_i and $v_{i \ominus 1}$ is *predecessor* of v_i. We call vertices $v_{i \oplus 2}$ and $v_{i \oplus 3}$ *opposite* to v_i.

Definition 1.

- P is a simple 5-gadget, *iff* $\forall v_i$, $|\angle v_{i \oplus 2} v_i v_{i \oplus 3}| \leq \frac{\pi}{2}$
- P is a blunt 5-gadget, *iff* $\exists v_i$, $|\angle v_{i \oplus 2} v_i v_{i \oplus 3}| > \frac{\pi}{2}$; *such angle is called a* blunt angle.

While we cannot avoid blunt 5-gadgets, at least we know that they have just one blunt angle:

Lemma 1. *A blunt 5-gadget has only one blunt angle.*

Proof. Assume there are two vertices v_i and v_j forming blunt angles. W.l.o.g., let v_0 be a vertex forming blunt angle with its opposite vertices.

If $i \ominus j = 1$, other vertex (named v_1) has to form blunt angle with vertices v_3 and v_4 (see Fig. 3 left). In order to maintain convexity, v_1 needs to be positioned

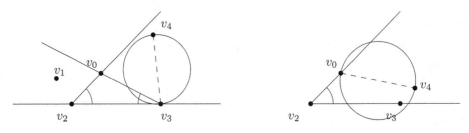

Fig. 3. At most one $> \frac{\pi}{2}$ angle with opposite vertices can be present in 5 vertices forming a convex hull. Left image describes case with two polar-adjacent blunt-angle vertices, right image two non-polar-adjacent blunt-angle vertices.

in sector defined by $\angle v_0 v_3 v_2$ and v_4 inside $\angle v_0 v_2 v_3$ and moreover, $\triangle v_0 v_2 v_3$ has to remain empty.

In order for v_1 to form angle $> \frac{\pi}{2}$, it has to be positioned inside circle of Tales, which has diameter defined by its two opposite vertices — v_3 and arbitrary vertex inside $\angle v_0 v_2 v_3 - \triangle v_0 v_2 v_3$. Area of this circle does not intersect area suitable for v_1.

If $i \ominus j > 1$, then in order to construct second blunt angle w.l.o.g at v_2, v_2's opposite vertices need to form diameter of circle of Tales, in which v_2 is positioned (see Fig. 3 right). Diameter is formed by fixed v_0 and variable v_4; however, r_4 lies in $\angle v_0 v_2 v_3$, but outside $\triangle v_0 v_2 v_3$, thus for any admittable v_4, the circle does not cross v_2 and thus is not possible to form blunt angle.

For 5 vertices, only two above-mentioned cases can occur, thus at most one vertex forming blunt angle with its opposite vertices can be present.

Definition 2. *Let $P = \{v_0, v_1 \ldots, v_4\}$ be a 5-gadget and let A be an antenna assignment on P.*

- *We say that A is* all-covering *iff every point of the plane is covered by an antenna assigned to some vertex of P.*
- *We say that A is* compact, *iff $D_h(IG(A, P)) \leq 2$.*
- *Let $r^*(v_i)$ denote $r(v_i) \cap \overline{CE(P)}$ and $l^*(v_i)$ denote $l(v) \cap \overline{CE(P)}$, where $\overline{CE(P)}$ is the complement of $CE(P)$[1].*
  We say that A is rotating, *iff $\forall i : v_{i \oplus 1}$ sees $r^*(v_i)$ and $v_{i \ominus 1}$ sees $l^*(v_i)$.*

Lemma 2. *Let P be a simple 5-gadget. Then there exists a compact, all-covering, rotating antenna assignment A for P.*

Proof. Consider the plane split by the lines connecting vertices of P into *main* sectors s_i and *side* sectors $s_{j,j \oplus 1}$ – see Fig. 2. Since P is a simple 5-gadget, each vertex is able to cover its two opposite vertices by an antenna, and thus the corresponding main sector can always be covered.

---

[1] Recall that $r(v)$ and $l(v)$ are the counter-clockwise and clockwise rays of antenna v, respectively.

Consider an algorithm assigning antennas in such way, that v_i's antenna covers $s_i \cup s_{i,i\oplus1}$. We call this the *default assignment*. If $s_{i,i\oplus1}$ can not be covered (the angle necessary to cover $s_i \cup s_{i,i\oplus1}$ from v_i is larger that $\pi/2$), we call it *uncoverable side-sector*.

We need to consider the following cases:

- **No uncoverable side-sector:** Use default assignment; as there are no uncoverable side sectors, this is an all-covering assignment. It is easy to verify that this is rotating assignment as well.
- **One uncoverable side-sector.** Assume there is one side-sector (w.l.o.g. $s_{2,3}$), which can not be fully covered by the default assignment.
  The antenna assignment is as follows:
  - use v_2's antenna to cover s_2 and as much of $s_{2,3}$ as possible (i.e. v_4v_2 is the clockwise boundary of its coverage)
  - use v_3's antenna to cover $s_{3,4}$, s_3 and as much of $s_{2,3}$ as possible (the counterclockwise boundary is parallel to v_4v_1)
  - use standard assignment for the remaining antennas
  Since only $s_{2,3}$ is uncoverable, in order to show that this construction is indeed all-covering, it is sufficient to show that the whole $s_{2,3}$ is covered. However, that follows from the fact that the sum of angles covered by v_2 and v_3 is $\pi - \angle v_2v_4v_1 < \pi$. From this and the fact that the remaining assignment is standard we obtain that this is a rotating assignment as well.
- **Two or more uncoverable side sectors.** Assume first that there are two uncoverable side-sectors next to each other in a simple 5-gadget (w.l.o.g. $s_{0,1}$ and $s_{1,2}$); thus, both v_0 and v_1 have sum of their respective main and side-sectors greater than $\frac{\pi}{2}$. Hence, $\alpha_1 = \angle v_2u_4v_3 > \frac{\pi}{2}$ and $\alpha_2 = \angle v_3u_0v_4 > \frac{\pi}{2}$. However, this is not possible as the triangle $u_4v_2u_0$ has α_1 and α_2 as its inner angles. Therefore, two uncoverable side-sectors are always separated by a coverable side-sector. Furthermore, it is not possible to have more than two uncoverable side-sectors, since at least two of them would not be separated by a coverable side-sector.
  The solution for the two uncoverable side sectors (w.l.o.g. assume $s_{2,3}$ and $s_{4,0}$) is now obtained by sequentially applying the solution for each such sector (i.e. v_2 and v_3 jointly cover $s_2 \cup s_{2,3} \cup s_3 \cup s_{3,4}$ and v_4 and v_0 jointly cover $s_4 \cup s_{4,0} \cup s_0 \cup s_{0,1}$).

Finally, from construction, $IG(A, P)$ contains cycle $v_0v_2v_4v_1v_3$, hence its diameter is at most 2 and the antenna assignment is compact.

Lemma 3. *Let P be a blunt 5-gadget. Then there exists a compact, all-covering and rotating antenna assignment A for P.*

Proof. W.l.o.g let v_0 be the vertex forming blunt angle with its two opposite vertices and let v_2v_3 be horizontal, with v_0 lying above them. From convexity we know that v_1 and v_4 can not be both higher than v_0. Hence, w.l.o.g assume that v_4 is lower than v_0. Antennas are oriented as follows: (see Fig. 4).

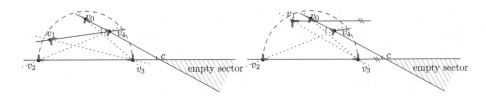

Fig. 4. Two of the cases when orienting the blunt-gadget's antennas.

- set antennas of v_0 and v_4 to see each other and cover lower halfplane of line v_0v_4
- set antennas of v_2 and v_3 to see each other and cover upper halfplane of line v_2v_3
- set antenna of v_1 to cover v_3 and v_4, and also the empty sector that is not covered by any of the halfplanes above.

We now show the following claims, which imply that A is all-covering and compact (the fact that it is rotating then follows from its construction):

(i) v_2 and v_4 see each other
(ii) v_0 and v_3 see each other
(iii) v_1 covers the empty sector and sees each other with v_3 and v_4.

v_0 forms blunt angle with vertices v_2 and v_3, therefore it lies inside circle of Tales defined by these two vertices. Statements (i) and (ii) follow from that, construction and from the fact that v_4 lies above v_0v_3 and to the right of v_0 (by convexity).

Since v_0 is inside the circle of Tales, $|\angle v_0v_3c| > \pi/2$ and therefore the angle of the empty sector is less than $\pi/2$. This, together with the fact that v_1 is inside the cone v_0cv_2, means that v_1 is able to cover the whole empty sector. In order to prove that v_1 also sees each other with v_3 and v_4, more detailed case analysis is necessary (see Fig. 4):

- Case $|\angle v_2cv_0| \leq |\angle v_2v_3v_1|$ and $|\angle v_0cv_2| \leq |\angle v_0v_4v_1|$: By covering v_3 and v_4, empty sector is always covered.
- Case $|\angle v_2cv_0| > |\angle v_2v_3v_1|$ and $|\angle v_0cv_2| \leq |\angle v_0v_4v_1|$: Orient antenna at v_1 in such way, that one of its edges is directed towards v_4 and the antenna covers the lower quadrant. By this, both v_3 and v_4 are covered. From the fact that v_0 is inside the circle of Tales and from convexity, the angle $v_1v_4v_0$ is always $\leq \frac{\pi}{2}$, thus the empty sector is covered as well.
- Case $|\angle v_2cv_0| \leq |\angle v_2v_3v_1|$ and $|\angle v_0cv_2| > |\angle v_0v_4v_1|$: Orient antenna at v_1 in such way, that one of its edges is directed towards v_3 and the antenna covers the upper quadrant. By this, both v_3 and v_4 are covered. By construction and convexity, $|\angle cv_3v_1 > \angle cv_3v_0 > \frac{\pi}{2}$, hence the empty sector is covered.
- Case $|\angle v_2cv_0| > |\angle v_2v_3v_1|$ and $|\angle v_0cv_2| > |\angle v_0v_4v_1|$: This means that positioning antenna's edge on a line between v_1 and one of v_3, v_4 is not sufficient. However, setting the antenna edge co-linear with v_2v_3 and covering the lower

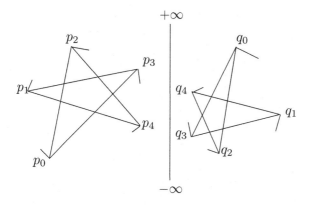

Fig. 5. Two 5-gadgets using default construction

quadrant makes it cover also the whole empty sector (as its angle is less than $\frac{\pi}{2}$). Furthermore, v_1 and v_3 (v_4) see each other due to the preconditions of this case (Fig. 5).

Lemma 4. *Let P and Q be two 5-gadgets separated by a line, and let A_P and A_Q be all-covering rotating antenna assignments on P and Q. Then there exist $p \in P$ and $q \in Q$ such that p and q see each other.*

Proof. W.l.o.g. we may assume l s vertical and P is to the left and Q is to the right of l, respectively. Let the vertices of P be labeled clockwise using labels p_0 through p_4; p_0 being the first vertex to cover the infinite top of l. Analogously, let the vertices of Q be labeled clockwise using labels q_0 through q_4; q_0 being the first one covering the infinite bottom of l. Note that there are vertices (inevitably those with largest indexes) that do not cover any part of l. Let p_{i*} and q_{j*} be the last vertices of P and Q, respectively, covering any part of l.

The fact that A_P and A_Q are rotating assignments allows us to define what it means for a vertex to be *after* some other vertex's antenna: Let the antennas of $p_i \in P$ and $q_j \in Q$ both cover some part of l. We say p_i *is after* q_j's antenna if p_i is not covered by q_j's antenna, and rotating q_j's antenna clockwise would cover p_i before it happens that q_j's antenna does not cover any part of l anymore. q_j being after p_i's antenna is defined analogously.

We are now ready to prove the lemma by contradiction. Assume there are no two vertices $p \in P$, $q \in Q$ such that they can see each other. Let us call a pair of vertices (p_i, q_j) *eligible*, iff one of them sees the other one, while the other one is after the first one's antenna.

From the fact that A_P and A_Q are all-covering, we know that an eligible pair must exist: Take p_0, all q_j's are either seen by p_0 or are after its antenna. It cannot be the case that all q_j's are seen by p_0, because in such case (from the fact that A_Q is all-covering) one of them would see p_0, contradicting the assumption that there are no two vertices from P and Q seeing each other.

Let (p_i, q_j) be the eligible pair for which $i + j$ is maximized.

W.l.o.g. assume p_i sees q_j and is after q_j's antenna. From the fact that A_Q is all-covering and rotating it follows that there exists $q_{j'}$ for some $j' > j$ such that $q_{j'}$ sees p_i. Since we assume no two vertices see each other, this means that p_i does not see $q_{j'}$. Futhermore, since A_P is rotating and from the way the vertices od Q are labeled, $q_{j'}$ cannot be before p_i's antenna, hence it must be after it. This means that $(p_i, q_{j'})$ is an eligible pair. However, $i + j' > i + j$, which contradicts the assumption that (p_i, q_j) be the eligible pair for which $i + j$ is maximized.

3   7-Spanner

The following algorithm is used to construct a maximal set $\mathcal{C}$ of 5-gadgets:

Algorithm 1. Selecting the 5-gadgets

$\mathcal{C} \leftarrow \emptyset; P' \leftarrow P;$
repeat
    Let L be a connected (in UDG) set of 9 vertices in P', such that
        $CE(L) \cap CE(C_i) = \emptyset$ for all $C_i \in \mathcal{C}$.
    Select (using Theorem 1) 5 vertices from L to form a 5-gadget,
        add the newly formed 5-gadget to $\mathcal{C}$ and remove its vertices from P'
until no such L can be found any more

Definition 3. *We say $v \in P$ is* free *vertex, iff for all $C_i \in \mathcal{C}$, $v \notin C_i$. Similarly, $v' \in P$ is called* core *vertex, iff $\exists C_i \in \mathcal{C}$, $v' \in C_i$.*

Observe that from construction, $\forall i, j, i \neq j : C_i \cap C_j = \emptyset$. Let $C(v)$ denote the closes (Euclidean distance) 5-gadget to a free vertex v.

Observation 1. *Let v be a free vertex. Then $d_e(v, CE(C(v))) \leq 8$.*

Proof. By contradiction, assume $d_e(v, CE(C(v))) > 8$. Consider the circle of radius 8 centered at v. Since UDG(S) is connected, there exists a path from v going outside the circle. The first 9 points (including v) of this path lie inside this circle. By assumption, the circle does not intersect $CE(C(v))$. However, that means that using Theorem 1, 5 vertices among those 9 should have been selected as a new 5-gadget, contradiction.

Note that the diameter of 5-gadgets is at most 8, as each of them is selected from a connected set of 9 vertices. Figure 6 shows that this is tight, i.e. it is possible to have a 5-gadget of radius $8 - \varepsilon$.

Observation 2. *Let $(u, v) \in$ UDG(S) such that $C(u) \neq C(v)$. Then $d_e(CE(C(u)), CE(C(v))) \leq 17$.*

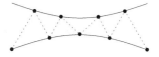

Fig. 6. Configuration of 9 vertices containing single convex pentagon. Flattening this configuration results in this pentagon having radius $8 - \varepsilon$.

Proof. By applying Observation 1 for u and v and counting the edge (u, v) we obtain

$$d_e(CE(C(u)), CE(C(v))) \leq d_e(CE(C(u), u)) + d_e(u, v) + d_e(u, CE(C(u))) \leq 8 + 1 + 8 = 17$$

Theorem 2. *Given a set of sensors S in 2D Euclidean plane such that $G = UDG(S)$ is connected, with each sensor equipped with one antenna of apex angle $\frac{\pi}{2}$ and radius 33, it is possible to construct antenna assignment on S such that $IG(S, A)$ is connected and is a 7-spanner of G.*

Proof. First, consider the case that $D_e(S) \leq 33$. In such case, apply the construction of Achner et. al. from [2] by constructing a 4-sensor hub using all-covering antenna assignment, and orient all the remaining vertices to point to a vertex in this hub that sees them. Since the hop-diameter of the hub is 3, the resulting stretch factor is 5.

In the rest of the proof, we assume $D_e(S) \geq 33$.

First, construct C using Algorithm 1. Then, use all-covering, compact, rotating antenna assignments from Sect. 2.2 for each 5-gadget $P \in C$. Finally, for each free vertex v, orient its antenna to the core vertex of $C(v)$ that sees v.

Let us show now that when using radius 33, the resulting induced graph is connected and has stretch factor 7.

Consider an arbitrary edge $(u, v) \in UDG(S)$:

- Case $C(u) = C(v) = C_i$: Since u and v communicate using the same 5-gadget, it takes at most one hop from u to C_i, at most 2 hops inside C_i and at most one hop from C_i to v. Stretch of the edge (u, v) is thus at most 5.

  From the fact that a free vertex is at most at distance 8 from its 5-gadget (Observation 1) and the fact that the diameter of a 5-gadget is at most 8 it follows that radius 16 is sufficient for these connections.
- Case $C_u \neq C_v$, both u and v are free vertices: The path from u to v now goes as follows:
  (a) from u to the vertex in $C(u)$ that sees it – 1 hop
  (b) to the vertex in $C(u)$ which is connected to a vertex in $C(v)$ (by Lemma 4 there must be such) – at most 2 hops (by compactness of the antenna assignment)
  (c) to $C(v)$ – 1 hop
  (d) to the vertex of $C(v)$ which is connected to v – at most 2 hops
  (e) to v – 1 hop.

The total number of hops (and hence the stretch factor) in this case is 7. Radius 8 and 16 are sufficient for steps (b), (d) and (a), (e), respectively. The largest radius is needed in step (c), connecting the two vertices of $C(u)$ and $C(v)$ that see each other – this is at most $D_e(C(u)) + d_e(CE(C(u)), CE(C(v))) + D_e(C(v)) \leq 8 + 17 + 8 = 33$, using Observation 2.

- Case $C_u \neq C_v$, but one or both of u and v are core vertices: The same construction as in the previous case will work, with the exception that the first and/or last hop is saved, resulting in stretch 5 or 6.

4   Stretch Factor 5

The stretch factor 7 in previous section is caused by edges (u, v), where $C(u) \neq C(v)$ and it is necessary to take two hops in both $C(u)$ and $C(v)$.

We use slightly modified antenna assignments and more careful analysis to isolate these problematic edges with stretch 7 into disjoint components of constant geometric diameter. We then reassign antennas to some of the free vertices to cover these segments efficiently.

In order for our construction to work, we need the 5-gadgets to be well-separated by a padding distance p (to be determined later), i.e. $\forall C_i, C_j \in \mathcal{C}$, $i \neq j : d_e(C_i, C_j) \geq p$ holds. This is achieved by modifying Algorithm 1 to use the following statement:

"Let L be a connected (in UDG) set of 9 vertices in P', such that $d_e(CE(L), CE(C_i)) \geq p$ for all $C_i \in \mathcal{C}$."

Let us denote by $D(C_i)$ the *domain* of the 5-gadget C_i. The domain of C_i starts as the set of points for which C_i is the closest 5-gadget, but it will be modified slightly later on. For a free vertex v, we can now define $C(v)$ to be the 5-gadget C_i, for which $v \in D(C_i)$.

Definition 4. *Let $(u, v) \in \mathrm{UDG}(S)$ such that $C(u) \neq C(v)$. Let u', v' be the vertices of $C(u)$ and $C(v)$, respectively, that see each other according to Lemma 4. If there are more such pairs, select one arbitrarily. We call the line segment $u'v'$ the* linking edge *for $C(u)$ and $C(v)$; the vertices u' and v' are called* linking vertices.

When given u and v, we will use u_v^* and v_u^*, respectively, to denote the linking vertices between $C(u)$ and $C(v)$. Note that there is a linking edge between two 5-gadgets C_i and C_j iff there is an edge $(u, v) \in \mathrm{UDG}(S)$ such that $C(u) = C_i$ and $C(v) = C_j$.

Before proceeding further, we need to slightly modify the domains of the 5-gadgets:

Observe that this process does not add new linking edges (some might be removed, though) as only already linked domains grow, therefore it eventually terminates. Furthermore, the domain of each 5-gadget did not shrink by more than a swath of width 2, hence they are still well-padded.

Because of the padding and the way the domains are defined, the analogue of Observation 1 now gives a larger possible distance from a free vertex to its 5-gadget:

Algorithm 2. Constructing the domains

1: Let $c(C_i)$ denote center of the smallest circle enclosing 5-gadget C_i.
2: Let $D(C_i)$ be the set of point such that $\forall x \in D(C_i), \forall j \neq i : d_e(c(C_i), x) \leq d_e(c(C_j), x)$.
3: **repeat**
4:     Let (u', v') be the linking edge between 5-gadgets C_i and C_j such that $D(C_i) \cap D(C_j) \neq \emptyset$.
5:         **if** $\exists l \notin \{i, j\}$ such that $X_l = \{x \in D(C_l) : d_e(x, (u', v')) \leq 2\} \neq \emptyset$ **then**
6:             Transfer the points of X_l from $D(C_l)$ to $D(C_i)$ and/or $D(C_j)$
7:                 (each point to the domain whose center is closer)
8:         **end if**
9: **until** No such X_l can be found any more.

Lemma 5. *Let v be a free vertex. Then $d_e(v, c(C(v))) \leq 14 + p$.*

Proof. We will start arguing using the original domains, and will add 2 to the bound due to possible adding of slice of width at most 2 by Algorithm 2.

Consider the circle of radius 8 centered at v and a path starting at v and leading out of this circle. As this path contains at least 9 vertices, if v's distance from the closest 5-gadget were at least $8 + p$, a new 5-gadget formed from these 9 vertices would have been added. Hence, the distance from v to the CE of its closest 5-gadget C_i is at most $8 + p$. Since the radius of the smallest circle enclosing C_i is at most 4, this yields $d_e(v, c(C_i)) \leq 12 + p$. It might be the case that $C_i \neq C(v)$, however $d_e(v, c(C(v))) < d_e(v, c(C_i))$ (otherwise C_i would be $C(v)$) and the statement of the lemma follows. □

Let us denote by $D^*(C_i)$ the convex envelope of the vertices lying in $D(C_i)$. While there is no apriori bound on the diameter of $D(C_i)$ (it is closely related to a Voronoi cell, after all), Lemma 5 tells us that $D_e(D^*(C_i)) \leq 28 + 2p$. Let us denote this bound by R_C, and let R'_C denote the maximal distance from a free vertex in $D(C_i)$ to a core vertex of C_i. Observe that $R'_C \leq d_e(v, c(C(v))) + 4$, as the radius of the smallest circle enclosing a 5-gadget is at most 4.

We are now ready to specify the antenna assignment (it will be refined later on to deal with the problematic edges):

R1: Construct $\mathcal{C}$ in such way so that $\forall C_i, C_j \in \mathcal{C}, i \neq j : d_e(CE(C_i), CE(C_j)) \geq p$
R2: Let v be a free vertex. Orient v's antenna towards a vertex of $C(v)$, which covers v.
R3: Let $(u, v) \in \text{UDG}(S)$. If $C(u) \neq C(v)$, u is not covered by u_v^*, v is not covered by v_u^*, but v is covered by u_v^*, v's antenna changes its orientation to point to u_v^*.

Let us more carefully identify the problematic edges:

Observation 3. *Let $(u, v) \in \text{UDG}(S)$. Even if $C(u) \neq C(v)$, if u is covered by u_v^* or v is covered by v_u^*, the stretch factor of (u, v) is at most 5.*

Proof. If u is covered by u_v^*, the path from u to v saves 2 hops by not crossing $C(u)$ but going from u to u_v^* and then immediately to v_u^*. An analogous argument applies if v is covered by v_u^*.

Definition 5. *We say that edge (u, v) is* problematic *iff $C(u) \neq C(v)$ and neither the precondition of Observation 3 nor the precondition of rule R3 applies to it. A vertex is* problematic *if it belongs to a problematic edge. A* problematic component *is a connected component in the graph induced in* UDG(S) *by the problematic vertices.*

Observe that if rule R3 has been applied to a vertex v, the resulting stretch factor of any edge (w, v) is at most 5: Because of the way the domains were modified, each neighbour w of v is either in $D(u)$ or in $D(v)$. If $C(w) = C(u)$, only the 5-gadget $C(u)$ needs to be crossed on the path from w to v. If $C(w) = C(v)$, $C(u)$ does not need to be crossed, as the path from w through $C(w) = C(v)$ arrives to $C(u)$ via the linking vertex u_v^* and continues directly to v.

The following lemma is crucial in proving that the problematic vertices can be grouped into disjoint problematic regions of constant size.

Lemma 6. *No problematic edge (u, v) crosses the linking edge between $C(u)$ and $C(v)$.*

Proof. Let (u, v) be an edge such that $C(u) \neq C(v)$, u is not covered by u_v^*, v is not covered by v_u^* (i.e. preconditions of Observation 3 are not satisfied) and (u, v) crosses the linking edge between $C(u)$ and $C(v)$. Because of the padding, the distance from u_v^* to u and v is larger than 2. Since the distance between u and v is at most 1, they straddle the linking edge and u_v^*'s antenna covers angle $\pi/2$ and the linking edge, either u or v must be covered by it. Since u is not, v must be, i.e. the preconditions of rule R3 are satisfied and (u, v) is not a problematic edge.

Lemma 6 together with the domain modification by Algorithm 2 ensures that no problematic component can cross a linking edge.

Lemma 7. *Let P be a problematic component. Then $D_e(P) \leq R_P = 162$.*

Proof. First, observe that a problematic vertex must lie within distance 1 from the boundary of the domain of its 5-gadget. These boundaries are (with exception of some areas near the linking edges affected by Algorithm 2) the bisectors separating the Voronoi cells of the centers of the 5-gadgets.

Since $R_C \leq 28 + 2p$, the only way for $D_e(P)$ to be larger is to include problematic vertices from the domains of multiple 5-gadgets. Figure 7 shows one such example, which might suggest that $D_e(P)$ could be unbounded.

The crucial observation is that if the angle α between the boundary lines of the domains of the neighbouring 5-gadgets is too small, the distance between the domain boundaries[2] remains too large and the problematic component does not

---

[2] It starts at least p, due to padding; the relevant place to look is at distance $14 + p$ from $c(C_i)$, as that it according to Lemma 5 the furthest place in $D(C_i)$ containing any vertices.

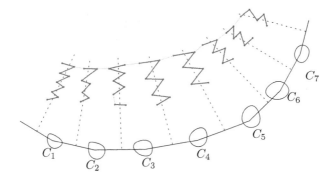

Fig. 7. Problematic component spreading like a plague. The dotted lines represent boundaries between the domains. The dots and the heavier lines connecting them are the problematic vertices and edges. The thin lines join problematic vertices at distance at most 1 to each other, connecting the problematic component across several domains.

connect to the problematic vertices of the next domain. Having a lower bound on α ensures that the diameter of the enclosed area in Fig. 7 is in fact constant.

Let us now compute the smallest α which still allows the problematic component to spread (refer to Fig. 8). In fact, it is not necessary to compute α, as bounding x from above directly leads to the bound on the diameter R_P.

Let us assume that the problematic component contains vertices in $D(C_{i'})$ and $D(C_{i''})$, connected through vertices of $D(C_i)$. For that to happen, the distance between points a and b must be at most 3 and these points must be (by Lemma 5) at distance at most $14 + p$ from $c(C_i)$. In fact, a more precise value is $12 + p$, as $+2$ is added only due to modifications by Algorithm 2 close to the linking edges – and a and b are far from those.

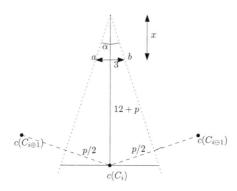

Fig. 8. Determining x and α.

Let us now determine x: From triangle similarity we get $\frac{x}{3} = \frac{12+p+x}{p}$, this yields $x(p-3) = 36 + 3p$ and hence $x = \frac{36+3p}{p-3}$.

Selecting $p = 26$ (that comes from other considerations) yields $x = \frac{114}{23} < 5$. The upper bound R_P on the diameter of P can now be calculated as (remember P does not cross outside the ring of the linking edges) $2(x + 12 + p) = 2x + 24 + 2p < 34 + 2p = 86$.

The *spare* vertices defined below will be used to efficiently handle the problematic components:

Definition 6. *Let spare vertices for C_i (denoted by $S(C_i)$) be a fixed independent set of five vertices, selected from free vertices in $D(C_i)$, such that no vertex from $S(C_i)$ is adjacent to a problematic vertex.*

Lemma 8. *If $p \geq 26$ then it is possible to construct $S(C_i)$ for each 5-gadget C_i.*

Proof. Since the distance between the 5-gadgets is at least p, the distance from $CE(C_i)$ to the boundary of $D(C_i)$ is at least $p/2 - 2$ (-2 comes from the way we readjusted the domains, possibly shrinking some). Since a problematic vertex can be only at distance at most 1 from the boundary of the domain, the free vertices inside $D(C_i)$ at distance at least than 2 from the domain's boundary are not adjacent to any problematic vertex. Let $P(C_i)$ be the shortest path from a core vertex of C_i towards the outside of $D(C_i)$. Since $p/2 - 4 \geq 9$, the first 9 (excluding the initial core one) vertices of $P(C_i)$ are free vertices and non-adjacent to a problematic vertex. As $P(C_i)$ is the shortest path, the only edges are among the neighbours in the path. This means the odd vertices of the path form an independent set of size 5, yielding $S(C_i)$.

Definition 7. *Let P be a problematic component. Problematic area $Q(P)$ is defined as $Q(P) = P \cup \bigcup_{v \in P} C(v)$*

Note that the diameter R_Q of $Q(P)$ is bound by $R_P + 2R'_C = 70 + 4p$. Selecting $p = 26$ yields $R_Q \leq 174$.

Definition 8. *An eligible 5-gadget for a problematic component P is any 5-gadget C_i such that $1.31R_Q < d_e(Q(P), D^*(C_i)) < 1.31R_Q + R'_C$.*

Note that $1/2\sin(\pi/8) < 1.31$; the lower bound on the distance for an eligible component has been chosen so that an antenna of angle $\pi/4$ placed anywhere in the domain of an eligible 5-gadget will be able to cover the whole $Q(P)$.

Observation 4. *If $D_e(S) \geq R_Q + 2(1.31R_Q + R'_C)$, every problematic component has an eligible 5-gadget.*

Proof. Consider a problematic component P. Since $D_e(S) \geq R_Q + 2(1.31R_Q + R'_C)$, there exists a vertex w at distance $1.31R_Q + R'_C$ from $Q(P)$. The 5-gadget $C(w)$ is the eligible 5-gadget for $Q(P)$.

The antenna assignment is completed by applying the following two rules:

R4: Let P be a problematic component and C_i be a fixed eligible 5-gadget for $Q(P)$. Orient the antennas of all vertices of P to cover $D(C_i)$.

R5: Let C_i be a gadget selected in rule R4 for some problematic component. Let $C_i = \{v_0, v_1, \ldots, v_4\}$ and $S(C_i) = \{s_0, s_1, \ldots, s_4\}$. Apply the following rule for $j = 0, \ldots 4$: Orient the antenna of s_j so that the center of its coverage is along the counter-clockwise ray $r(a_j)$ of v_j's antenna. If the antenna of s_j does not fully cover any problematic area that selected C_i in rule R4, revert this orientation (making s_j a free vertex again).

Lemma 9. *Let C_i be the eligible 5-gadget selected in rule R4 for problematic component P. Then either $Q(P)$ is covered by an antenna of a core vertex of C_i, or it is covered by an antenna of a spare vertex of C_i.*

Proof. Since C_i is eligible for $Q(P)$, an antenna of angle $pi/4$ starting at any point of $D^*(C_i)$ is able to cover $Q(P)$. If $Q(P)$ is not covered by an antenna of a single core vertex, this means that there are core vertices v_j and $v_{j\oplus 1}$ such that their antennas partially cover $Q(P)$. However, in such case the antenna of s_i covers the whole $Q(P)$.

Theorem 3. *Given a set of sensors S in 2D Euclidean plane such that $G = UDG(S)$ is connected, with each sensor equipped with one antenna of apex angle $\frac{\pi}{2}$ and radius 718, it is possible to construct antenna assignment on S such that $IG(S, A)$ is connected and is a 7-spanner of G.*

Proof. If $D_e(S) < R_Q + 2(1.31R_Q + R'_C)$, the approach from [2] can be used: Construct just one 4-sensor hub and direct all remaining vertices to it. The stretch factor is at most 5 for any edge (u, v): (1 hop from u to the hub, 3 hops within hub, 1 hop to v). The radius required is $R_Q + 2(1.31R_Q + R'_C) < 718$.

In the case $D_e(S) \geq R_Q + 2(1.31R_Q + R'_C)$, the antenna assignment is constructed using rules R1, R2, $\ldots$, R5, by Lemma 4 this is always possible. From the discussion at the beginning of this section and from Observation 3 it follows that the only edges which might have stretch larger than 5 are the ones incident to spare or problematic vertices.

Let us analyze those:

- Both u and v are problematic: Then they both belong to the same problematic component, both point to the same eligible 5-gadget and both are covered by the same (core or spare) vertex w. The stretch factor is therefore 2.
- u is problematic, v is normal (i.e. free or core): Let P be the problematic component u belongs to. Let w be the (core or spare) vertex covering $Q(P)$ and let w' be the vertex of $C(v)$ covering v. By definition, $C(v) \in Q(P)$ and therefore w covers a vertex w'' of $C(v)$ that sees it. The path from u to v then takes 1 hop from u to w, 1 hop from w to w'', at most 2 hops within $C(v)$ to reach w' and final 1 hop from w' to v. The resulting stretch is hence 5.
- u is spare, v is normal: From the construction in rule R5, u is spare only if there are indeed problematic vertices that selected $C(u)$ in rule R4 and u covers them. Let w be a problematic vertex that sees a core vertex w' of $C(u)$. The path from u to v then goes from u to w (1 hop), to w' (1 hop), to the vertex seeing v (at most 2 hops), to v; altogether 5 hop. (Note that from the construction of the spare vertices, $C(u) = C(v)$ in this case.)

Since spare vertices are chosen so that they do not neighbour another spare vertex, nor a problematic vertex, these are the only possibilities.

Note that in this case, the radius $R_Q + 1.31R_Q + R'_C < 446$ is sufficient to cover the longest communication link (connecting a core/problematic vertex of R_Q to the spare/core vertex of the corresponding eligible 5-gadget.)

5   Conclusions

The results presented in this paper are a significant improvement over the previous related results w.r.t. the best achievable stretch factor in the considered setting. In fact, further decrease of the stretch factor is not feasible using the approach of hubs and nearby free vertices, as it is not possible to have hubs of diameter 1.

However, this does not mean that our bounds are the best achievable. In particular, the following questions remain open:

- What is the true lower bound for the stretch factor using constant radius and antennas of spread $\pi/2$? A straightforward lower bound is 3 (for antennas of spread up to $\pi - \varepsilon$, with points on a line). Is it possible to improve the stretch below 5, at least for spread sharply less than π?
- In this paper we have not attempted to minimize the radius required for stretch factor 5. Significant improvement should be possible, the question is whether it can be pushed all the way down to a reasonable value. There is also space to improve the radius for stretch factor 7.

References

1. Aschner, R., Katz, M.J.: Bounded-angle spanning tree: modeling networks with angular constraints. In: Esparza, J., Fraigniaud, P., Husfeldt, T., Koutsoupias, E. (eds.) ICALP 2014, Part II. LNCS, vol. 8573, pp. 387–398. Springer, Heidelberg (2014)
2. Aschner, R., Katz, M.J., Morgenstern, G.: Symmetric connectivity with directional antennas. Comput. Geom. 46(9), 1017–1026 (2013)
3. Bose, P., Carmi, P., Damian, M., Flatland, R., Katz, M.J., Maheshwari, A.: Switching to directional antennas with constant increase in radius and hop distance. In: Dehne, F., Iacono, J., Sack, J.-R. (eds.) WADS 2011. LNCS, vol. 6844, pp. 134–146. Springer, Heidelberg (2011)
4. Caragiannis, I., Kaklamanis, C., Kranakis, E., Krizanc, D., Wiese, A.: Communication in wireless networks with directional antennas. In: Proceedings of the 20th Annual Symposium on Parallelism in Algorithms and Architectures (SPAA), pp. 344–351. ACM (2008)
5. Carmi, P., Katz, M.J., Lotker, Z., Rosen, A.: Connectivity guarantees for wireless networks with directional antennas. Comput. Geom. Theor. Appl. 44(9), 477–485 (2011)
6. Damian, M., Flatland, R.: Spanning properties of graphs induced by directional antennas. In: Electronic Proceedings of 20th Fall Workshop on Computational Geometry. Stony Brook, NY (2010)

7. Dobrev, S., Eftekhari, M., MacQuarrie, F., Manuch, J., Morales Ponce, O., Narayanan, L., Opatrny, J., Stacho, L.: Connectivity with directional antennas in the symmetric communication model. In: Mexican Conference on Discrete Mathematics and Computational Geometry. Oaxaca, Mexico (2013)
8. Dobrev, S., Kranakis, E., Krizanc, D., Morales-Ponce, O., Opatrny, J., Stacho, L.: Strong connectivity in sensor networks with given number of directional antennae of bounded angle. Discret. Math. Algorithms Appl. **4**(03), 72–86 (2012)
9. Eftekhari Hesari, M., Kranakis, E., MacQuarie, F., Morales-Ponce, O., Narayanan, L.: Strong connectivity of sensor networks with double antennae. In: Even, G., Halldórsson, M.M. (eds.) SIROCCO 2012. LNCS, vol. 7355, pp. 99–110. Springer, Heidelberg (2012)
10. Kranakis, E., Krizanc, D., Morales, O.: Maintaining connectivity in sensor networks using directional antennae. In: Nikoletseas, S., Rolim, J. (eds.) Theoretical Aspects of Distributed Computing in Sensor Networks. Springer, Heidelberg (2010)
11. Kranakis, E., MacQuarrie, F., Morales-Ponce, O.: Stretch factor in wireless sensor networks with directional antennae. In: Lin, G. (ed.) COCOA 2012. LNCS, vol. 7402, pp. 25–36. Springer, Heidelberg (2012)
12. Morris, W., Soltan, V.: The Erdős-Szekeres problem on points in convex position— a survey. Bull. Am. Math. Soc. New Ser. **37**(4), 437–458 (2000)
13. Shepard, C., Yu, H., Anand, N., Li, E., Marzetta, T., Yang, R., Zhong, L.: Argos: Practical many-antenna base stations. In: Proceedings of the 18th Annual International Conference on Mobile computing and Networking, pp. 53–64. ACM (2012)

Wireless Networks

Exploiting Geometry in the SINR_k Model

Rom Aschner[1], Gui Citovsky[2], and Matthew J. Katz[1(✉)]

[1] Department of Computer Science, Ben-Gurion University, Beer-Sheva, Israel
{romas,matya}@cs.bgu.ac.il
[2] Department of Applied Mathematics and Statistics, Stony Brook University,
Stony Brook, NY, USA
gui.citovsky@stonybrook.edu

Abstract. We introduce the SINR_k model, which is a practical version of the SINR model. In the SINR_k model, in order to determine whether s's signal is received at c, where s is a sender and c is a receiver, one only considers the k most significant senders w.r.t. to c (other than s). Assuming uniform power, these are the k closest senders to c (other than s). Under this model, we consider the well-studied scheduling problem: Given a set L of sender-receiver requests, find a partition of L into a minimum number of subsets (rounds), such that in each subset all requests can be satisfied simultaneously. We present an $O(1)$-approximation algorithm for the scheduling problem (under the SINR_k model). For comparison, the best known approximation ratio under the SINR model is $O(\log n)$. We also present an $O(1)$-approximation algorithm for the maximum capacity problem (i.e., for the single round problem), obtaining a constant of approximation which is considerably better than those obtained under the SINR model. Finally, for the special case where $k = 1$, we present a PTAS for the maximum capacity problem. Our algorithms are based on geometric analysis of the SINR_k model.

1 Introduction

The *SINR* (Signal to Interference plus Noise Ratio) model has received a lot of attention in recent years. It is considered a more realistic model for the behavior of a wireless network than the common graph-based models such as the unit disk graph, since it takes into account physical parameters such as the fading of the signal, interference caused by other transmitters and ambient noise. A fundamental problem in this context is the following: Given a set of communication sender-receiver requests, find a good scheduling for the requests. In other words, what is the minimum number of rounds needed to satisfy all the requests, such that in each round some subset of the communication links is active?

More formally, let $L = \{(c_1, s_1), (c_2, s_2), \ldots, (c_n, s_n)\}$ be a set of n pairs of points in the plane representing n (directional) links, where the points $c_1, \ldots, c_n$

Work by R. Aschner was partially supported by the Lynn and William Frankel Center for Computer Sciences. Work by R. Aschner, G. Citovsky, and M. Katz was partially supported by grant 2010074 from the United States – Israel Binational Science Foundation. Work by M. Katz was partially supported by grant 1045/10 from the Israel Science Foundation.

© Springer-Verlag Berlin Heidelberg 2015
J. Gao et al. (Eds.): ALGOSENSORS 2014, LNCS 8847, pp. 125–135, 2015.
DOI: 10.1007/978-3-662-46018-4_8

represent the receivers and the points $s_1, \ldots, s_n$ represent the senders. The *length* of the link $(c_i, s_i) \in L$ is the Euclidean distance between c_i and s_i (i.e., $|c_i s_i|$) and is denoted l_i. We denote the Euclidean distance between c_i and s_j, for $j \neq i$, by l_{ij}. The set of all receivers is denoted $C = C(L)$ and the set of all senders is denoted $S = S(L)$. Finally, let p_i be the transmission power of sender s_i, for $i = 1, \ldots, n$. In the SINR model, a link (c_i, s_i) is *feasible*, if c_i receives the signal sent by s_i. That is, if the following inequality holds (assuming all senders in S are active):

$$\frac{p_i/l_i^\alpha}{\sum\limits_{\{j:s_j \in S \setminus \{s_i\}\}} p_j/l_{ij}^\alpha + N} \geq \beta \,,$$

where $\alpha, \beta \geq 1$ and $N > 0$ are appropriate constants (α is the path-loss exponent, N is the ambient noise, and β is the threshold above which a signal is received successfully).

The *scheduling problem* is thus to partition the set of links L to a minimum number of *feasible* subsets (i.e., rounds), where a subset L_i is *feasible* if, when only the senders in $S(L_i)$ are active, each of the links in L_i is feasible. A greedy algorithm that successively finds a feasible subset of maximum cardinality of the yet unscheduled links yields an $O(\log n)$-approximation. Therefore, it is interesting to first focus on the *maximum capacity* problem, i.e., find a feasible subset of L of maximum cardinality. In other words, find a set $Q \subseteq L$, such that if only the senders in $S(Q)$ are active, then each of the links in Q is feasible, and Q is of maximum cardinality.

In the SINR model, the affectance of senders that are close to a receiver is much more significant than the affectance of those that are far from it. Moreover, in many scenarios the interference at a receiver is caused by a few nearby senders, while signals from farther senders are drastically degraded by, e.g., walls and distance. This has led us to define a restricted but more practical version of the SINR model which we name $SINR_k$.

The $SINR_k$ model. In this model, in order to determine whether a link (c, s) is feasible, one only considers the k most significant senders w.r.t. to c (other than s), which are the k closest senders to c (other than s) assuming uniform power. Formally, for a receiver c_i, let S_i^k be the set of the k most significant senders w.r.t. to c_i (other than s_i). Then, the link (c_i, s_i) is *feasible* if the following inequality holds (assuming all senders in S are active):

$$\frac{p_i/l_i^\alpha}{\sum\limits_{\{j:s_j \in S_i^k\}} p_j/l_{ij}^\alpha + N} \geq \beta \,.$$

Assuming uniform power, we examined the validity of the $SINR_k$ model in the specific but common setting where the senders are located on an $m \times m$ grid, for some odd integer m. Specifically, consider the sender s located at the center of the grid (i.e., at location $((m + 1)/2, (m + 1)/2)$). Let R denote the *reception region* of s; i.e., the region consisting of all points in the plane at which

Table 1. The ratio area(R_k)/area(R) for several values of k, computed for a sender at the center of a 31×31 grid.

k	area(R_k)/area(R)
4	1.102
8	1.039
12	1.029
20	1.017
24	1.014
28	1.012
36	1.011
44	1.006
$31^2 - 1$	1

s is received according to the SINR inequality (assuming all senders are active). Avin et al. [1] showed that R is convex and fat. Let R_k denote the reception region of s according to the SINR$_k$ inequality, i.e., when only the k closest neighbors of s are taken into account. Notice that for any two positive integers k_1, k_2, if $k_1 < k_2$, then $R_{k_1} \supset R_{k_2}$. We thus computed the region R_k for several values of k, and observed the rate at which R_k's area decreases as k increases. Consider Table 1 and Fig. 1. In this example, $m = 31$ (that is, we have 961 senders), $\alpha = 4$, and $\beta = 2$. The values in the left column are those for which we computed R_k, and the values in the right column are the corresponding ratios between the area of R_k and the area of R. Notice that already for $k = 4$, R_k's area is larger than R's area by only roughly 10 %, and that for $k = 44$ the difference drops to roughly 0.5 %; see Fig. 1.

Related work. The pioneering work of Gupta and Kumar [6] has initiated an extensive study of the maximum capacity and the scheduling problems in the SINR model. Several versions of these problem have been considered, depending on the capabilities of the underlying hardware, that is, whether and to what extent one can control the transmission power of the senders.

For the case where the transmission powers are given, Goussevskaia et al. [5] showed that the maximum capacity and the scheduling problems are NP-complete, even for uniform power. They also presented an $O(g(L))$-approximation algorithm, assuming uniform power, for the (weighted) maximum capacity problem, where $g(L)$ is the so-called diversity of the network, which can be arbitrarily large in general. Assuming uniform power, Chafekar et al. [2] presented an $O(\log \Delta)$-approximation algorithm for the maximum capacity problem, where Δ is the ratio between the longest link and the shortest link. If the ratio between the maximum power and the minimum power is bounded by Γ, then they give an $O(\log \Delta \log \Gamma)$-approximation algorithm for the problem. Goussevskaia et al. [4] and Halldórsson and Wattenhofer [9] gave constant-factor approximation

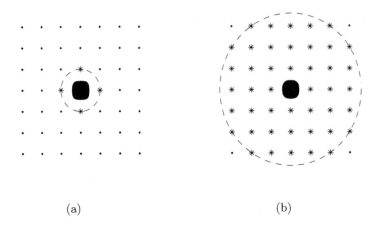

Fig. 1. The 7×7 neighborhood of a sender s located at the center of a 31×31 grid. (a) R_4, the reception region of s for $k = 4$ (the black spot around the center). (b) R_{44}, the reception region of s for $k = 44$ (the black spot around the center).

algorithms for the maximum capacity problem yielding an $O(\log n)$-approximation algorithm for the scheduling problem, assuming uniform power. In [4] they note that their $O(1)$-approximation algorithm also applies to the case where the ratio between the maximum power and the minimum power is bounded by a constant and for the case where the number of different power levels is constant. Later, Wan et al. [13] presented a constant-factor approximation algorithm for the maximum capacity problem, assuming uniform power; their constant is significantly better than the one in [4]. Recently, Halldórsson and Mitra [8] have considered the case of oblivious power. This is a special case of non-uniform power where the power of a link is a simple function of the link's length. They gave an $O(1)$-approximation algorithm for the maximum capacity problem, yielding an $O(\log n)$-approximation algorithm for scheduling. Finally, the case with (full) power control has also been studied, see, e.g., [7,8,10,12].

Our results. We study the maximum capacity and scheduling problems in the SINR_k model, for a given constant k, under the common assumptions that (i) $p_i = p_j$, for $1 \le i, j \le n$, i.e., uniform power (see, e.g., [4,13]), and (ii) $N = 0$, i.e., there is no ambient noise (see, e.g., [5]). We exploit some of the geometric properties of the SINR_k model to obtain $O(1)$-approximation algorithms for both problems. For comparison, the best known approximation ratio for the scheduling problem in the SINR model is $O(\log n)$. We also consider a variant of the maximum capacity problem in which one is free to form the links, and the goal, as in the standard problem, is to find a maximum-cardinality feasible subset of links. We obtain an $O(1)$-approximation algorithm for this variant as well. Finally, for $k = 1$, we present a PTAS for the maximum capacity problem.

The paper is organized as follows: In Sect. 2, we prove several geometric properties of the SINR_k model and use them to obtain an $O(1)$-approximation algorithm for the maximum capacity problem, where the constant that we get is significantly better than the one in [4]. In Sect. 3, we present an $O(1)$-approximation algorithm for the scheduling problem. Finally, in Sect. 4, we show that in the special case where $k = 1$, one can obtain a PTAS for the capacity problem, by using a technique due to Chan [3] that is based on geometric separators. To the best of our knowledge our work is the first to study the SINR$_k$ model.

2 Maximum Capacity

Let k be a positive integer. In this section we consider the *maximum capacity* problem under the SINR$_k$ model, assuming uniform power and no ambient noise. W.l.o.g., we shall assume that the transmission power of each of the senders is 1. Let $L' \subseteq L$ and (c_i, s_i) a link in L'. We say that (c_i, s_i) is *feasible* (in L'), if c_i receives s_i when only the senders in $S(L')$ are active. If all the links in L' are feasible, then we say that L' is *feasible*. Our goal is to find a feasible subset of links of maximum cardinality. We begin by proving a series of lemmas establishing several important geometric properties of feasible links.

Lemma 1. *Let $L' \subseteq L$ and let $(c_i, s_i) \in L'$ be a feasible link. Then, the disk centered at c_i of radius $\sqrt[\alpha]{\beta} \cdot l_i$ does not contain in its interior any sender of $S(L')$ except for s_i.*

Proof. Assume that this disk contains another sender (except s_i) in its interior. Let s_r be such a sender, i.e., $l_{ir} < \sqrt[\alpha]{\beta} l_i$. Then

$$\frac{1/l_i^\alpha}{\Sigma_{\{j:s_j \in S_i^k\}} 1/l_{ij}^\alpha} \leq \frac{1/l_i^\alpha}{1/l_{ir}^\alpha} < \beta,$$

where $S_i^k \subseteq S(L')$ is the set of the k closest senders to c_i, not including s_i. This is a contradiction to the assumption that (c_i, s_i) is a feasible link in L'.

Lemma 2. *Let $L' \subseteq L$ and let $(c_i, s_i), (c_j, s_j) \in L'$ be two feasible links. Let $D_i = D(c_i, m \cdot l_i)$, $D_j = D(c_j, m \cdot l_j)$ be two disks around the two receivers. If $m < \frac{\sqrt[\alpha]{\beta}-1}{2}$, then $D_i \cap D_j = \emptyset$.*

Proof. By Lemma 1, $l_{ij} \geq \sqrt[\alpha]{\beta} \cdot l_i$ and $l_{ji} \geq \sqrt[\alpha]{\beta} \cdot l_j$, that is, $l_i + l_j \leq \frac{1}{\sqrt[\alpha]{\beta}}(l_{ij} + l_{ji})$. By the triangle inequality, $l_{ij} \leq |c_i c_j| + l_j$ and $l_{ji} \leq |c_j c_i| + l_i$, and therefore $l_i + l_j \leq \frac{1}{\sqrt[\alpha]{\beta}}(2|c_i c_j| + l_i + l_j)$. Rearranging, we get that $|c_i c_j| \geq \frac{\sqrt[\alpha]{\beta}-1}{2}(l_i + l_j) > m(l_i + l_j) = ml_i + ml_j$. This implies that, $D_i \cap D_j = \emptyset$.

The following lemma is actually a generalization of Lemma 1.

Lemma 3. *Let $L' \subseteq L$ and let $(c_i, s_i) \in L'$ be a feasible link. Then, the disk centered at c_i of radius $\sqrt[\alpha]{\beta k} \cdot l_i$ contains in its interior at most $k - 1$ senders of $S(L') \setminus \{s_i\}$.*

Proof.

$$\beta \leq \frac{1/l_i^\alpha}{\Sigma_{\{j:s_j \in S_i^k\}} 1/l_{ij}^\alpha} \leq \frac{1/l_i^\alpha}{k \min_{\{j:s_j \in S_i^k\}} \{1/l_{ij}{}^\alpha\}} \leq \frac{\max_{\{j:s_j \in S_i^k\}} \{l_{ij}^\alpha\}}{k l_i^\alpha}.$$

Thus, $\sqrt[\alpha]{\beta k} \cdot l_i \leq \max_{\{j:s_j \in S_i^k\}} \{l_{ij}\}$. That is, the farthest among the k senders in S_i^k does not lie in the interior of the disk centered at c_i of radius $\sqrt[\alpha]{\beta k} \cdot l_i$.

Lemma 4. *Let $L' \subseteq L$ and let $(c_i, s_i) \in L'$. If the disk centered at c_i of radius $\sqrt[\alpha]{\beta k} \cdot l_i$ does not contain in its interior any sender of $S(L')$ (except for s_i), then the link (c_i, s_i) is feasible.*

Proof. For each $s_j \in S_i^k$, we have that $l_{ij} \geq \sqrt[\alpha]{\beta k} \cdot l_i$. Therefore, $\Sigma_{\{j:s_j \in S_i^k\}} \frac{1}{l_{ij}^\alpha} \leq \frac{1}{\beta l_i^\alpha}$, and $\frac{1/l_i^\alpha}{\Sigma_{\{j:s_j \in S_i^k\}} 1/l_{ij}^\alpha} \geq \beta$.

2.1   An $O(1)$-Approximation for Constant k

For each $(c_i, s_i) \in L$, let D_i denote the disk of radius $\sqrt[\alpha]{\beta k} \cdot l_i$ centered at c_i, and set $\mathcal{D} = \{D_i | (c_i, s_i) \in L\}$.

We apply the following simple (and well-known) algorithm that finds an independent set $\mathcal{Q}$ in the intersection graph induced by $\mathcal{D}$, such that $|\mathcal{Q}|$ is at least some constant fraction of the size of a maximum independent set in this graph. We then prove that the set of links corresponding to $\mathcal{Q}$ is an $O(1)$-approximation of OPT, where OPT is an optimal solution for the maximum capacity problem (under SINR_k). This proof is non-trivial since the disks in $\mathcal{D}$ corresponding to the links in OPT are not necessarily disjoint.

Algorithm 1. An $O(1)$-approximation

$\quad \mathcal{Q} \leftarrow \emptyset$
Sort $\mathcal{D}$ by the radii of the disks in increasing order.
while $\mathcal{D} \neq \emptyset$ **do**
$\quad$ Let D be the smallest disk in $\mathcal{D}$
$\quad \mathcal{D} \leftarrow \mathcal{D} \setminus \{D\}$
$\quad$ **for all** $D' \in \mathcal{D}$, such that $D \cap D' \neq \emptyset$ **do**
$\quad\quad \mathcal{D} \leftarrow \mathcal{D} \setminus \{D'\}$
$\quad \mathcal{Q} \leftarrow \mathcal{Q} \cup \{D\}$
return $\mathcal{Q}$

Algorithm 1 returns a subset $\mathcal{Q} \subseteq \mathcal{D}$ which is an independent set, i.e., for any two disks $D_1, D_2 \in \mathcal{Q}$, $D_1 \cap D_2 = \emptyset$. Moreover, by Lemma 4, the subset of links corresponding to $\mathcal{Q}$ is feasible. From now on, we shall mostly think of OPT as a set of disks, i.e., the subset of disks in $\mathcal{D}$ corresponding to the links in OPT. Below we show that $|OPT| = O(|\mathcal{Q}|)$.

Lemma 5. *Let $L' \subseteq L$ be a feasible set of links, and let $D(L')$ denote the set of corresponding disks of radius $\sqrt[\alpha]{\beta k} \cdot l_i$ around the receivers c_i in $C(L')$. Then, every point $p \in \mathbb{R}^2$ is covered by at most $\tau = \dfrac{2\pi(k+1)}{\arctan(\frac{\sqrt[\alpha]{\beta k}-1}{\sqrt[\alpha]{\beta k}+1})}$ disks in $D(L')$.*

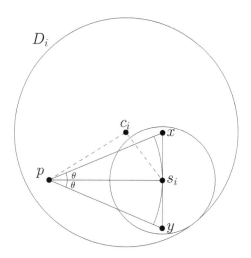

Fig. 2. Proof of Lemma 5.

Proof. Let $p \in \mathbb{R}^2$, and consider the set $D(L_p) \subseteq D(L')$ of all disks in $D(L')$ that cover p. Let s_i be the sender (among the senders in $S(L_p)$) that is farthest from p. We draw a wedge W of angle 2θ and apex p, where $\theta = \arctan(\frac{\sqrt[\alpha]{\beta k}-1}{\sqrt[\alpha]{\beta k}+1})$, such that s_i is on its bisector (see Fig. 2). We claim that the disk D_i (of radius $\sqrt[\alpha]{\beta k} \cdot l_i$ and center c_i) covers all the senders in $S(L_p) \cap W$.

Consider the line perpendicular to $\overline{ps_i}$ and passing through s_i, and let x and y be the intersection points of this line with W's rays. Then,

$$|xs_i| = |ps_i|\tan(\theta) \leq (|pc_i| + l_i)\tan(\theta) \leq (\sqrt[\alpha]{\beta k} + 1)l_i\tan(\theta) = (\sqrt[\alpha]{\beta k} - 1)l_i \ .$$

Similarly, $|ys_i| \leq (\sqrt[\alpha]{\beta k} - 1)l_i$. Therefore the disk of radius $(\sqrt[\alpha]{\beta k} - 1)l_i$ and center s_i contains points x and y. But this disk is contained in the disk D_i. So, D_i contains the triangle $\triangle pxy$ (since it covers its three corners), and, since all the senders in $S(L_p) \cap W$ lie in $\triangle pxy$, we conclude that D_i covers all these senders. This implies that the number of these senders is at most $k+1$.

We now remove all the senders in $S(L_p) \cap W$ and repeat. After at most $2\pi/\theta$ iterations, we finish removing all senders in $S(L_p)$. Thus, the number of senders in $S(L_p)$ is at most τ, implying that the number of disks in $D(L')$ covering p is at most τ.

Lemma 6. $|OPT| \leq 8\tau|\mathcal{Q}|$.

Proof. First notice that each disk in OPT intersects at least one of the disks in Q. Since, otherwise, consider the smallest disk in OPT that does not intersect any of the disks in Q. Then, our algorithm would have chosen this disk – contradiction. We thus associate each disk in OPT with the smallest disk in Q which it intersects. Let $D = D(c, r) \in Q$. We show that the number of disks associated with D is at most 8τ. We first observe that each of the disks associated with D is at least as large as D. Since, if one or more of these disks were smaller than D, then our algorithm would have chosen the smallest of them instead of D – contradiction.

Let A be a set of 8 points including (i) the center point c, and (ii) seven points evenly spaced on a circle of radius $3r/2$ around c; see Fig. 3. Notice that any disk that is not smaller than D and intersects D must cover at least one of the points in A. In particular, this is true for each of the disks in OPT associated with D. By Lemma 5, there are at most τ disks in OPT covering each of these points. Thus, at most 8τ disks in OPT intersect D. We conclude that our algorithm computes a $(1/8\tau)$-approximation of OPT.

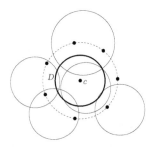

Fig. 3. Any disk that is not smaller than D and intersects D covers at least one of the 8 points.

The following theorem summarizes the main result of this section.

Theorem 1. *Given a set L of n links and a constant k, one can compute a $(1/8\tau)$-approximation for the maximum capacity problem under the $SINR_k$ model, where $\tau = \dfrac{2\pi(k+1)}{\arctan(\frac{\sqrt[6]{\beta k}-1}{\sqrt[6]{\beta k}+1})}$.*

2.2   All Pairs Maximum Capacity

We now consider the maximum capacity problem where any sender and receiver can be paired. Let $L = \{(c_i, s_j) | 1 \leq i, j \leq n\}$ be a set of n^2 potential links. We seek a feasible subset of links $Q \subseteq L$ of maximum cardinality, enforcing a one-to-one correspondence between $S(Q)$ and $C(Q)$.

For each $(c_i, s_j) \in L$, let D_{ij} denote the disk of radius $\sqrt[6]{\beta k} \cdot l_{ij}$ centered at c_i (where $l_{ii} = l_i$), and set $\mathcal{D} = \{D_{ij} | (c_i, s_j) \in L\}$. We apply Algorithm 1 with $\mathcal{D}$ as our input set of disks. Note that any time a pair (c_i, s_j) is added to Q, all other potential links in L that contain c_i as a receiver or s_j as a sender will be removed from consideration. This is because all other disks in $\mathcal{D}$ either using c_i as receiver or s_j as a sender clearly have a nonempty intersection with D_{ij}. Lemma 6 shows that Algorithm 1 gives an $O(1)$-approximation for the all pairs version as well. Namely, $|OPT| \leq 8\tau|\mathcal{Q}|$.

3   Scheduling

In this section we consider the scheduling problem. That is, given a set L of links (i.e., requests), how many rounds are needed to satisfy all the requests? Alternatively, find a partition of L into a minimum number of feasible subsets.

We show how to obtain a constant factor approximation for the scheduling problem under the SINR$_k$ model. As in the previous section, for each $(c_i, s_i) \in L$, let D_i denote the disk of radius $\sqrt[\alpha]{\beta k} \cdot l_i$ centered at c_i, and set $\mathcal{D} = \{D_i | (c_i, s_i) \in L\}$. The *depth* of a point $p \in \mathbb{R}^2$ with respect to $\mathcal{D}$ is the number of disks in $\mathcal{D}$ covering p. The *depth* of $\mathcal{D}$ is the depth of a point $p \in \mathbb{R}^2$ of maximum depth (i.e., it is the depth of the arrangement of the disks in $\mathcal{D}$). Notice that the depth of $\mathcal{D}$ is not necessarily bounded.

Let r be the number of rounds in an optimal solution, OPT, to the scheduling problem. We first observe that the depth of $\mathcal{D}$ is $O(r)$.

Lemma 7. *The depth of $\mathcal{D}$ is $O(r)$, where r is the number of rounds in OPT.*

Proof. Let L_i be the set of active links in round i, for $1 \leq i \leq r$. By Lemma 5 every point in the plane is covered by at most τ disks in $D(L_i)$ (i.e., the depth of $D(L_i)$ is at most τ). Therefore, the depth of $D(L)$ is at most τr.

Miller et al. [11] showed how to color an intersection graph of a set of balls in $\mathbb{R}^d$ of bounded ply. In particular, their result implies a polynomial-time algorithm for coloring the intersection graph of the disks in $\mathcal{D}$ with $9\tau r + 1$ colors. Each color class is an independent set, and thus, by Lemma 4, is a feasible solution.

The following theorem summarizes the main result of this section.

Theorem 2. *Given a set L of n links and a constant k, one can compute a $(9\tau + 1)$-approximation for the scheduling problem under the SINR$_k$ model.*

4   A PTAS for Maximum Capacity with $k = 1$

By plugging $k = 1$ in Lemmas 3 and 4, we obtain the following lemma.

Lemma 8. *Let $L' \subseteq L$ and let $(c_i, s_i) \in L'$. Then, (c_i, s_i) is a feasible link if and only if the disk centered at c_i of radius $\sqrt[\alpha]{\beta} \cdot l_i$ does not contain in its interior any sender of $S(L')$ (except for s_i).*

The following theorem is due to Timothy Chan [3].

Theorem 3 ([3]). *Given a measure μ satisfying the following five conditions, a collection $\mathcal{O}$ of n objects in $\mathbb{R}^d$ and $\varepsilon > 0$, one can find a $(1+\varepsilon)$-approximation to $\mu(\mathcal{O})$ in $O(n^{O(1/\varepsilon^d)})$ time and $O(n)$ space.*

1. *If $\mathcal{A} \subseteq \mathcal{B}$, then $\mu(\mathcal{A}) \le \mu(\mathcal{B})$.*
2. *$\mu(\mathcal{A} \cup \mathcal{B}) \le \mu(\mathcal{A}) + \mu(\mathcal{B})$.*
3. *If for any pair $(A, B) \in \mathcal{A} \times \mathcal{B}$, $A \cap B = \emptyset$, then $\mu(\mathcal{A} \cup \mathcal{B}) = \mu(\mathcal{A}) + \mu(\mathcal{B})$.*
4. *Given any r and size-r box R, if every object in $\mathcal{A}$ intersects R and has size at least r, then $\mu(\mathcal{A}) \le c$ for a constant c.*
5. *A constant-factor approximation to $\mu(\mathcal{A})$ can be computed in time $|\mathcal{A}|^{O(1)}$. If $\mu(\mathcal{A}) \le b$, then $\mu(\mathcal{A})$ can be computed exactly in time $|\mathcal{A}|^{O(b)}$ and linear space.*

Chan has applied this theorem to the measures pack$(\cdot)$ and pierce$(\cdot)$ and a collection of fat objects. We apply this theorem in a somewhat non-standard manner to obtain our PTAS.

For each $(c_i, s_i) \in L$, let D_i denote the disk of radius $\sqrt[\alpha]{\beta} \cdot l_i$ centered at c_i, and set $\mathcal{D} = \{D_i | (c_i, s_i) \in L\}$. For any $\mathcal{A} \subseteq \mathcal{D}$, let $\mu(\mathcal{A})$ denote the cardinality of a feasible subset of $\mathcal{A}$ of maximum cardinality. Below, we show that μ satisfies the five conditions above.

Notice first that two disks D_1 and D_2 in a feasible subset $\mathcal{D}'$ of $\mathcal{D}$ may intersect; in particular, one or both of the receivers c_1, c_2 may lie in the other disk. However, none of the senders s_1, s_2 may lie in the other disk. Conditions (1) and (2) are clearly satisfied. Condition (3) is also satisfied, since the assumption implies that none of the senders corresponding to the disks in $\mathcal{B}$ lies in a disk of $\mathcal{A}$ and vise versa. Concerning Condition (4), we can apply Lemma 5 in a similar way to the one described in the proof of Lemma 6, to show that (under the assumption of Condition (4)) $\mu(\mathcal{A})$ is bounded by some constant. Finally, Algorithm 1 computes a constant-factor approximation to $\mu(\mathcal{A})$ in time $|\mathcal{A}|^{O(1)}$.

The following theorem summarizes the main result of this section.

Theorem 4. *Given a set L of n links and $\varepsilon > 0$, one can compute a $(1-\varepsilon)$-approximation for the maximum capacity problem under the $SINR_1$ model.*

Remark. Notice that the only condition that is not satisfied when k is a constant greater than 1, is Condition (3). The reason for this is that for $k > 1$ the converse of Lemmas 3 is no longer true.

References

1. Avin, C., Emek, Y., Kantor, E., Lotker, Z., Peleg, D., Roditty, L.: SINR diagrams: convexity and its applications in wireless networks. J. ACM **59**(4), 18 (2012)
2. Chafekar, D., Kumar, V.S.A., Marathe, M.V., Parthasarathy, S., Srinivasan, A.: Approximation algorithms for computing capacity of wireless networks with SINR constraints. In: INFOCOM, pp. 1166–1174 (2008)
3. Chan, T.M.: Polynomial-time approximation schemes for packing and piercing fat objects. J. Algorithms **46**(2), 178–189 (2003)

4. Goussevskaia, O., Halldórsson, M.M., Wattenhofer, R., Welzl, E.: Capacity of arbitrary wireless networks. In: INFOCOM, pp. 1872–1880 (2009)
5. Goussevskaia, O., Oswald, Y.A., Wattenhofer, R.: Complexity in geometric SINR. In: MobiHoc, pp. 100–109 (2007)
6. Gupta, P., Kumar, P.R.: The capacity of wireless networks. IEEE Trans. Inf. Theory 46(2), 388–404 (2000)
7. Halldórsson, M.M.: Wireless scheduling with power control. ACM Trans. Algorithms 9(1), 7 (2012)
8. Halldórsson, M.M., Mitra, P.: Wireless capacity with oblivious power in general metrics. In: SODA, pp. 1538–1548 (2011)
9. Halldórsson, M.M., Wattenhofer, R.: Wireless communication is in APX. In: Albers, S., Marchetti-Spaccamela, A., Matias, Y., Nikoletseas, S., Thomas, W. (eds.) ICALP 2009, Part I. LNCS, vol. 5555, pp. 525–536. Springer, Heidelberg (2009)
10. Kesselheim, T.: A constant-factor approximation for wireless capacity maximization with power control in the SINR model. In: SODA, pp. 1549–1559 (2011)
11. Miller, G.L., Teng, S.-H., Thurston, W.P., Vavasis, S.A.: Separators for sphere-packings and nearest neighbor graphs. J. ACM 44(1), 1–29 (1997)
12. Moscibroda, T., Wattenhofer, R.: The complexity of connectivity in wireless networks. In: INFOCOM (2006)
13. Wan, P.-J., Jia, X., Yao, F.: Maximum independent set of links under physical interference model. In: Liu, B., Bestavros, A., Du, D.-Z., Wang, J. (eds.) WASA 2009. LNCS, vol. 5682, pp. 169–178. Springer, Heidelberg (2009)

Interference Minimization
in Asymmetric Sensor Networks

Yves Brise[1], Kevin Buchin[2], Dustin Eversmann[3],
Michael Hoffmann[1], and Wolfgang Mulzer[3]([✉])

[1] ETH Zürich, Zurich, Switzerland
hoffmann@inf.ethz.ch
[2] TU Eindhoven, Eindhoven, The Netherlands
k.a.buchin@tue.de
[3] FU Berlin, Berlin, Germany
mulzer@inf.fu-berlin.de

Abstract. A fundamental problem in wireless sensor networks is to connect a given set of sensors while minimizing the *receiver interference*. This is modeled as follows: each sensor node corresponds to a point in $\mathbb{R}^d$ and each *transmission range* corresponds to a ball. The receiver interference of a sensor node is defined as the number of transmission ranges it lies in. Our goal is to choose transmission radii that minimize the maximum interference while maintaining a strongly connected asymmetric communication graph.

For the two-dimensional case, we show that it is NP-complete to decide whether one can achieve a receiver interference of at most 5. In the one-dimensional case, we prove that there are optimal solutions with nontrivial structural properties. These properties can be exploited to obtain an exact algorithm that runs in quasi-polynomial time. This generalizes a result by Tan et al. to the asymmetric case.

1 Introduction

Wireless sensor networks constitute a popular paradigm in mobile networks: several small independent devices are distributed in a certain region, and each device has limited computational resources. The devices can communicate through a wireless network. Since battery life is limited, it is imperative that the overhead for the communication be kept as small as possible. Thus, the literature on sensor networks contains many strategies to reduce the number of communication links while maintaining desirable properties of the communication networks. The term *topology control* refers to the general paradigm of dropping edges from the communication graph in order to decrease energy consumption.

Traditionally, topology control focused on properties such as sparsity, dilation, or congestion of the communication graph. This changed with the work of

KB supported in part by the Netherlands Organisation for Scientific Research (NWO) under project no. 612.001.207. WM supported in part by DFG Grants MU 3501/1 and MU 3502/2.

J. Gao et al. (Eds.): ALGOSENSORS 2014, LNCS 8847, pp. 136–151, 2015.
DOI: 10.1007/978-3-662-46018-4_9

Burkhart et al. [3], who pointed out the importance of explicitly considering the *interference* caused by competing senders. By reducing the interference, we can avoid costly retransmission of data due to data collisions, leading to increased battery life. At the same time, we need to ensure that the resulting communication graph remains connected.

There are many different ways to formalize the problem of interference minimization [4,6,8,9,11]. Usually, the devices are modeled as points in d-dimensional space, and the transmission ranges are modeled as d-dimensional balls. Each point can choose the radius of its transmission range, and different choices of transmission ranges lead to different reachability structures. There are two ways to interpret the resulting communication graph. In the *symmetric* case, the communication graph is undirected, and it contains an edge between two points p and q if and only if both p and q lie in the transmission range of the other point [6,8,9,11]. For a valid assignment of transmission ranges, we require that the communication graph is connected. In the *asymmetric* case, the communication graph is directed, and there is an edge from p to q if and only if p lies in the transmission range of q. We require that the communication graph is strongly connected, or, in a slightly different model, that there is one point that is reachable from every other point through a directed path [4].

In both the symmetric and the asymmetric case, the (*receiver-centric*) *interference* of a point is defined as the number of transmission ranges that it lies in [12]. The goal is to find a valid assignment of transmission ranges that makes the maximum interference as small as possible. We refer to the resulting interference as *minimum interference*. The minimum interference under the two models for the asymmetric case differs by at most one: if there is a point reachable from every other, we can increase its transmission range to include all other points. As a result, the communication graph becomes strongly connected, while the minimum interference increases by at most one. All of these models have been also considered in a non-euclidean setting, in which the problems studied in this paper cannot be approximated efficiently under standard assumptions [1].

Let n be the number of points. In the symmetric case, one can always achieve interference $O(\sqrt{n})$, and this is sometimes necessary [5,12]. In the one-dimensional case, there is an efficient approximation algorithm with approximation factor $O(n^{1/4})$ [12]. Furthermore, Tan et al. [13] prove the existence of optimal solutions with interesting structural properties in one dimension. This can be used to obtain a nontrivial exact algorithm for this case. In the asymmetric case, the interference is significantly smaller: one can always achieve interference $O(\log n)$, which is sometimes optimal (e.g., [7]). The one-dimensional model is also called the *highway model* [11]. Rickenbach et al. [11] cite the "intuition that already one-dimensional networks exhibit most of the complexity of finding minimum-interference topologies" as their motivation to study this model. Based on our experiences with the model, we can only support this claim.

There exist several other interference models. For example, the *sender-centric* model of Burkhart et al. [3] aims to minimize the maximum interference caused by any *edge* of the communication graph (i.e., the total number of points that lie in a certain region defined by the edge). Moscibroda and Wattenhofer [9] present a more general model that works in abstract metric spaces and that can distinguish between active and passive nodes. Johannsson and Carr-Motyčková [6] define a notion of *average path interference*, where the interference is taken as the average of the interferences over all interference-optimal paths in the network.

Our results. We consider interference minimization in asymmetric wireless sensor networks in one and two dimensions. We show that for two dimensions, it is NP-complete to find a valid assignment that minimizes the maximum interference. In one dimension we consider our second model requiring one point that is reachable from every other point through a directed path. Generalizing the result by Tan et al. [13], we show that there is an optimal solution that exhibits a certain binary tree structure. By means of dynamic programming, this structure can be leveraged for a nontrivial exact algorithm. Unlike the symmetric case, this algorithm always runs in quasi-polynomial time $2^{O(\log^2 n)}$, making it unlikely that the one-dimensional problem is NP-hard. Nonetheless, a polynomial time algorithm remains elusive.

2   Preliminaries and Notation

We now formalize our interference model for the planar case. Let $P \subset \mathbb{R}^2$ be a planar n-point set. A *receiver assignment* $N : P \to P$ is a function that assigns to each point in P the furthest point that receives data from P. The resulting (asymmetric) *communication graph* $G_P(N)$ is the directed graph with vertex set P and edge set $E_P(N) = \{(p, q) \mid \|p - q\| \leq \|p - N(p)\|\}$, i.e., from each point $p \in P$ there are edges to all points that are at least as close as the assigned receiver $N(p)$. The receiver assignment N is *valid* if $G_P(N)$ is strongly connected.

For $p \in \mathbb{R}^2$ and $r > 0$, let $B(p, r)$ denote the closed disk with center p and radius r. We define $B_P(N) = \{B(p, d(p, N(p))) \mid p \in P\}$ as the set that contains for each $p \in P$ a disk with center p and $N(p)$ on the boundary. The disks in $B_P(N)$ are called the *transmission ranges* for N. The *interference* of N, $I(N)$, is the maximum number of transmission ranges that cover a point in P, i.e., $I(N) = \max_{p \in P} |\{p \in B \mid B \in B_P(N)\}|$. In the *interference minimization problem*, we are looking for a valid receiver assignment with minimum interference.

3   NP-completeness in Two Dimensions

We show that the following problem is NP-complete: given a planar point set P, does there exist a valid receiver assignment N for P with $I(N) \leq 5$? It follows that the minimum interference for planar point sets is NP-hard to approximate within a factor of 6/5.

The problem is clearly in NP. To show that interference minimization is NP-hard, we reduce from the problem of deciding whether a grid graph of maximum degree 3 contains a Hamiltonian path: a *grid graph* G is a graph whose vertex set

$V \subset \mathbb{Z} \times \mathbb{Z}$ is a finite subset of the integer grid. Two vertices $u, v \in V$ are adjacent in G if and only if $\|u - v\|_1 = 1$, i.e., if u and v are neighbors in the integer grid. A *Hamiltonian path* in G is a path that visits every vertex in V exactly once. Papadimitriou and Vazirani showed that it is NP-complete to decide whether a grid graph G of maximum degree 3 contains a Hamiltonian cycle [10]. Note that we may assume that G is connected; otherwise there can be no Hamiltonian path.

Our reduction proceeds by replacing each vertex v of the given grid graph G by a *vertex gadget* P_v; see Fig. 1. The vertex gadget consists of 13 points, and it has five parts: (a) the *main point* M with the same coordinates as v; (b) three *satellite stations* with two points each: $S_1, S_1', S_2, S_2', S_3, S_3'$. The coordinates of the S_i are chosen from $\{v \pm (0, 1/4), v \pm (1/4, 0)\}$ so that there is a satellite station for each edge in G that is incident to v. If v has degree two, the third satellite station can be placed in any of the two remaining directions. The S_i' lie at the corresponding clockwise positions from $\{v \pm (\varepsilon, 1/4), v \pm (1/4, -\varepsilon)\}$, for a sufficiently small $\varepsilon > 0$; (c) the *connector* C, a point that lies roughly at the remaining position from $\{v \pm (0, 1/4), v \pm (1/4, 0)\}$ that is not occupied by a satellite station, but an ε-unit further away from M. For example, if $v + (0, 1/4)$ has no satellite station, then C lies at $v + (0, 1/4 + \varepsilon)$; and (d) the *inhibitor*, consisting of five points $I_c, I_1, \ldots, I_4$. The point I_c is the center of the inhibitor and I_1 is the point closest to C. The position of I_c is $M + 2(C - M) + \varepsilon(C - M)/\|C - M\|$, that is, the distance between I_c and C is an ε-unit larger than the distance between C and M: $\|M - C\| + \varepsilon = \|C - I_c\|$. The points $I_1, \ldots, I_4$ are placed at the positions $\{I_c \pm (0, \varepsilon), I_c \pm (\varepsilon, 0)\}$, with I_1 closest to C.

Given a grid graph G, the reduction can be carried out in polynomial time: just replace each vertex v of G by the corresponding gadget P_v; see Fig. 2 for an example. Let $P = \bigcup_{v \in G} P_v$ be the resulting point set. Two satellite stations in P that correspond to the same edge of G are called *partners*. First, we investigate the interference in any valid receiver assignment for P.

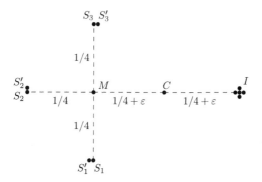

Fig. 1. The vertex gadget.

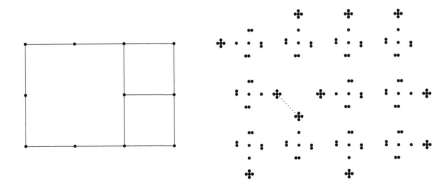

Fig. 2. An example reduction.

Lemma 3.1. *Let N be a valid receiver assignment for P. Then in each vertex gadget, the points I_c and M have interference as least 5, and the points S_1, S_2, and S_3 have interference at least 3.*

Proof. For each point $p \in P$, the transmission range $B(p, d(p, N(p)))$ must contain at least the nearest neighbor of p. Furthermore, in each satellite station and in each inhibitor, at least one point must have an assigned receiver outside of the satellite station or inhibitor; otherwise, the communication graph $G_P(N)$ would not be strongly connected. This forces interference of 5 at M and at I_c: each satellite station and C must have an edge to M, and $I_1, \ldots, I_4$ all must have an edge to I_c. Similarly, for $i = 1, \ldots 3$, the main point M and the satellite S'_i must have an edge to S_i; see Fig. 3.   □

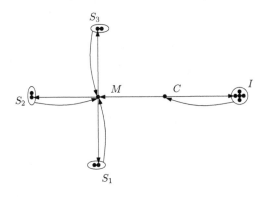

Fig. 3. The nearest neighbors in a vertex gadget.

Let N be a valid receiver assignment, and let P_v be a vertex gadget in P. An *outgoing* edge for P_v is an edge in $G_P(N)$ that originates in P_v and ends in a different vertex gadget. An *incoming* edge for P_v is an edge that originates in a different gadget and ends in P_v. A *connecting* edge for P_v is either an outgoing or an incoming edge for P_v. If $I(N) \leq 5$ holds, then Lemma 3.1 implies that a connecting edge can be incident only to satellite stations.

Lemma 3.2. *Let N be a valid receiver assignment for P with $I(N) \leq 5$. Let P_v be a vertex gadget of P and e an outgoing edge from P_v to another vertex gadget P_w. Then e goes from a satellite station of P_v to its partner satellite station in P_w. Furthermore, in each satellite station of P_v, at most one point is incident to outgoing edges.*

Proof. By Lemma 3.1, both M and I_c in P_v have interference at least 5. This implies that neither M, nor C, nor any point in the inhibitor of P_v can be incident to an outgoing edge of P_v: such an edge would increase the interference at M or at I_c. In particular, note that the distance between the inhibitors in two distinct vertex gadgets is at least $\sqrt{2}/2 - O(\varepsilon) > 1/2 + O(\varepsilon)$, the distance between M and its corresponding inhibitor; see the dotted line in Fig. 2.

Thus, all outgoing edges for P_v must originate in a satellite station. If there were a satellite station in P_v where both points are incident to outgoing edges, the interference at M would increase. Furthermore, if there were a satellite station in P_v with an outgoing edge that does not go the partner station, this would increase the interference at the main point of the partner vertex gadget, or at the inhibitor center I_v of P_v. □

Next, we show that the edges between the vertex gadgets are quite restricted.

Lemma 3.3. *Let N be a valid receiver assignment for P with $I(N) \leq 5$. For every vertex gadget P_v in P, at most two satellite stations in P_v are incident to connecting edges in $G_P(N)$.*

Proof. By Lemma 3.2 connecting edges are between satellite stations and by Lemma 3.1, the satellite points S_i in P_v have interference at least 3.

First, assume that all three satellite stations in P_v have outgoing edges. This would increase the interference at all three S_i to 5. Then, P_v could not have any incoming edge from another vertex gadget, because this would increase the interference for at least one S_i (note that due to the placement of the S_i', every incoming edge causes interference at an S_i). If P_v had no incoming edge, $G_P(N)$ would not be strongly connected. It follows that P_v has at most two satellite stations with outgoing edges.

Next, assume that two satellite stations in P_v have outgoing edges. Then, the third satellite station of P_v cannot have an incoming edge, as the two outgoing edges already increase the interference at the third satellite station to 5.

Hence, we know that every vertex gadget P_v either (i) has connecting edges with all three partner gadgets, exactly one of which is outgoing, or (ii) is connected to at most two other vertex gadgets. Take a vertex gadget P_v of type (i) with partners P_{u_1}, P_{u_2}, P_w. Suppose that P_v has incoming edges from P_{u_1} and P_{u_2} and that the outgoing edge goes to P_w. Follow the outgoing edge to P_w. If P_w is of type (i), follow the outgoing edge from P_w; if P_w is of type (ii) and has an outgoing edge to a vertex gadget we have not seen yet, follow this edge.

Continue this process until P_v is reached again or until the next vertex gadget has been visited already. This gives all vertex gadgets that are reachable from P_v on a directed path. However, in each step there is only one choice for the next vertex gadget. Thus, the process cannot discover P_{u_1} and P_{u_2}, since both of them would lead to P_v in the next step, causing the process to stop. It follows that at least one of P_{u_1} or P_{u_2} is not reachable from P_v, although $G_P(N)$ should be strongly connected. Therefore, all vertex gadgets in $G_P(N)$ must be of type (ii), as claimed in the lemma.   □

We can now prove the main theorem of this section.

Theorem 3.4. *Given a point set $P \subset \mathbb{R}^2$, it is NP-complete to decide whether there exists a valid receiver assignment N for P with $I(N) \le 5$.*

Proof. Using the receiver assignment N as certificate, the problem is easily seen to be in NP. To show NP-hardness, we use the polynomial time reduction from the Hamiltonian path problem in grid graphs: given a grid graph G of maximum degree 3, we construct a planar point set P as above. It remains to verify that G has a Hamiltonian path if and only if P has a valid receiver assignment N with $I(N) \le 5$.

   Given a Hamilton path H in G, we construct a valid receiver assignment N for P as follows: in each vertex gadget, we set $N(M) = C$, $N(C) = M$, and $N(I_1) = C$. For $i = 1, \ldots, 3$ we set $N(S_i') = S_i$ and $N(I_{i+1}) = I_c$. Finally, we set $N(I_c) = I_1$. This essentially creates the edges from Fig. 3, plus the edge from M to C. Next, we encode H into N: for each S_i on an edge of H, we set $N(S_i)$ to the corresponding S_i in the partner station. For the remaining S_i, we set $N(S_i) = M$. Since H is Hamiltonian, $G_P(N)$ is strongly connected (note that each vertex gadget induces a strongly connected subgraph). It can now be verified that M and I_c have interference 5; I_2, I_3, I_4 have interference 2; and I_1 has interference 3. The point C has interference between 2 and 4, depending on whether S_1 and S_3 are on edges of H. The satellites S_i and S_i' have interference at most 5 and 4, respectively.

   Now consider a valid receiver assignment N for P with $I(N) \le 5$. Let F be the set of edges in G that correspond to pairs of vertex gadgets with a connecting edge in $G_P(N)$. Let H be the subgraph that F induces in G. By Lemma 3.3, H has maximum degree 2. Furthermore, since $G_P(N)$ is strongly connected, the graph H is connected and meets all vertices of G. Thus, H is a Hamiltonian path (or cycle) for G, as desired.   □

Remark. A similar result to Theorem 3.4 also holds for symmetric communication graphs networks [2].

4   The One-Dimensional Case

For the one-dimensional case we minimize receiver interference under the second model discussed in the introduction: given $P \subset \mathbb{R}$ and a receiver assignment

$N : P \to P$, the graph $G_P(N)$ now has a directed edge from each point $p \in P$ to its assigned receiver $N(p)$, and no other edges. N is *valid* if $G_P(N)$ is acyclic and if there is a sink $r \in P$ that is reachable from every point in P. The sink has no outgoing edge. The interference of N, $I(N)$, is defined as before.

4.1   Properties of Optimal Solutions

We now explore the structure of optimal receiver assignments. Let $P \subset \mathbb{R}$ and N be a valid receiver assignment for P with sink r. We can interpret $G_P(N)$ as a directed tree, so we call r the *root* of $G_P(N)$. For a directed edge pq in $G_P(N)$, we say that p is a *child* of q and q is the *parent* of p. We write $p \rightsquigarrow_N q$ if there is a directed path from p to q in $G_P(N)$. If $p \rightsquigarrow_N q$, then q is an *ancestor* of p and p a *descendant* of q. Note that p is both an ancestor and a descendant of p. Two points $p, q \in P$ are *unrelated* if p is neither an ancestor nor a descendant of q. For two points p, q, we define $((p,q)) = (\min\{p,q\}, \max\{p,q\})$ as the open interval bounded by p and q, and $[[p,q]] = [\min\{p,q\}, \max\{p,q\}]$ as the closure of $((p,q))$. An edge pq of $G_P(N)$ is a *cross edge* if the interval $((p,q))$ contains at least one point that is not a descendant of p.

Our main structural result is that there is always an optimal receiver assignment for P without cross edges. A similar property was observed by Tan et al. for the symmetric case [13].

Lemma 4.1. *Let N^* be a valid receiver assignment for P with minimum interference. There is a valid assignment $\widetilde{N}$ for P with $I(\widetilde{N}) = I(N^*)$ such that $G_P(\widetilde{N})$ has no cross edges.*

Proof. Pick a valid assignment $\widetilde{N}$ with minimum interference that minimizes the total length of the cross edges

$$C(\widetilde{N}) := \sum_{pq \in C} \|p - q\|,$$

where C are the cross-edges of $G_P(\widetilde{N})$. If $C(\widetilde{N}) = 0$, we are done. Thus, suppose $C(\widetilde{N}) > 0$. Pick a cross edge pq such that the hop-distance (i.e., the number of edges) from p to the root is maximum among all cross edges. Let p_l be the leftmost and p_r the rightmost descendant of p.

Proposition 4.2. *The interval $[p_l, p_r]$ contains only descendants of p.*

Proof. Since p_l and p_r each have a path to p, the interval $[p_l, p_r]$ is covered by edges that begin in proper descendants of p. Thus, if $[p_l, p_r]$ contains a point z that is not a descendant of p, then z would be covered by an edge $p_1 p_2$ with p_1 a proper descendant of p. Thus, $p_1 p_2$ would be a cross edge with larger hop-distance to the root, despite the choice of pq. $\qquad\square$

Let R be the points in $((p,q))$ that are not descendants of p. Each point in R is either unrelated to p, or it is an ancestor of p. Let $z \in R$ be the point in R that is closest to p (i.e., z either lies directly to the left of p_l or directly to the right of p_r). We now describe how to construct a new valid assignment $\widehat{N}$, from which we will eventually derive a contradiction to the choice of $\widetilde{N}$. The construction is as follows: replace the edge pq by pz. Furthermore, if (i) $q \leadsto_{\widetilde{N}} z$; (ii) the last edge $z'z$ on the path from q to z crosses the interval $[p_l, p_r]$; and (iii) $z'z$ is not a cross-edge, we also change the edge $z'z$ to the edge that connects z' to the closer of p_l or p_r.

Proposition 4.3. $\widehat{N}$ *is a valid assignment.*

Proof. We must show that all points in $G_P(\widehat{N})$ can reach the root. At most two edges change: pq and (potentially) $z'z$. First, consider the change of pq to pz. This affects only the descendants of p. Since z is not a descendant of p, the path from z to the root does not use the edge pq, and hence all descendants of p can still reach the root. Second, consider the change of $z'z$ to an edge from z' to p_l or p_r. Both p_l and p_r have z as ancestor (since we introduced the edge pz), so all descendants of z' can still reach the root. $\square$

Proposition 4.4. *We have* $I(N^*) = I(\widehat{N})$.

Proof. Since the new edges are shorter than the edges they replace, each transmission range for $\widehat{N}$ is contained in the corresponding transmission range for $\widetilde{N}$. The interference cannot decrease since N^* is optimal. $\square$

Proposition 4.5. *We have* $C(\widehat{N}) < C(\widetilde{N})$.

Proof. First, we claim that $\widehat{N}$ contains no new cross edges, except possibly pz: suppose ab is a cross edge of $G_P(\widehat{N})$, but not of $G_P(\widetilde{N})$. This means that $((a, b))$ contains a point x with $x \leadsto_{\widetilde{N}} a$, but $x \not\leadsto_{\widehat{N}} a$. Then x must be a descendant of p in $G_P(\widetilde{N})$ and in $G_P(\widehat{N})$, because as we saw in the proof of Claim 4.3, for any $y \in P \setminus [p_l, p_r]$, we have that if $y \leadsto_{\widetilde{N}} a$, then $y \leadsto_{\widehat{N}} a$.

Hence, $((a, b))$ and $[p_l, p_r]$ intersect. Since ab is a cross edge, the choice of pq now implies that $[p_l, p_r] \subseteq ((a, b))$. Thus, z lies in $[[a, b]]$, because z is a direct neighbor of p_l or p_r. We claim that $b = z$. Indeed, otherwise we would have $z \leadsto_{\widetilde{N}} a$ (since ab is not a cross edge in $G_P(\widehat{N})$), and thus also $z \leadsto_{\widehat{N}} a$. However, we already observed $x \leadsto_{\widehat{N}} p$, so we would have $x \leadsto_{\widehat{N}} a$ (recall that we introduce the edge pz in $\widehat{N}$). This contradicts our choice of x.

Now it follows that $ab = az$ is the last edge on the path from p to z, because if a were not an ancestor of p, then ab would already be a cross-edge in $G_P(\widetilde{N})$. Hence, (i) a is an ancestor of q; (ii) az crosses the interval $[p_l, p_r]$; and (iii) az is not a cross edge in $\widetilde{N}$. These are the conditions for the edge $z'z$ that we remove from $\widetilde{N}$. The new edge e from a to p_l or p_r cannot be a cross edge, because ab is not a cross edge in $G_P(\widehat{N})$ and e does not cover any descendants of p.

Hence, $G_P(\widehat{N})$ contain no new cross-edges, except possibly pz which replaces the cross edge pq. By construction, $\|p - z\| < \|p - q\|$, so $C(\widehat{N}) < C(\widetilde{N})$. $\square$

Propositions 4.3–4.5 yield a contradiction to the choice of $\widetilde{N}$. It follows that we must have $C(\widetilde{N}) = 0$, as desired.   □

Let $P \subset \mathbb{R}$. We say that a valid assignment N for P has the *BST-property* if the following holds for any vertex p of $G_P(N)$: (i) p has at most one child q with $p < q$ and at most one child q with $p > q$; and (ii) let p_l be the leftmost and p_r the rightmost descendant of p. Then $[p_l, p_r]$ contains only descendants of p. In other words: $G_P(N)$ constitutes a binary search tree for the (coordinates of the) points in P. A valid assignment without cross edges has the BST-property. The following is therefore an immediate consequence of Lemma 4.1.

Theorem 4.6. *Every $P \subset \mathbb{R}$ has an optimal valid assignment with the BST-property.*   □

4.2   A Quasi-Polynomial Algorithm

We now show how to use Theorem 4.6 for a quasi-polynomial time algorithm to minimize the interference. The algorithm uses dynamic programming. A subproblem π for the dynamic program consists of four parts: (i) an interval $P_\pi \subseteq P$ of *consecutive* points in P; (ii) a root $r_\pi \in P_\pi$; (iii) a set I_π of *incoming interference*; and (iv) a set O_π of *outgoing interference*.

The objective of π is to find an optimal valid assignment N for P_π subject to (i) the root of $G_N(P_\pi)$ is r; (ii) the set O_π contains all transmission ranges of $B_{P_\pi}(N)$ that cover points in $P \setminus P_\pi$ plus potentially a transmission range with center r_π; (iii) the set I_π contains transmission ranges that cover points in P_π and have their center in $P \setminus P_\pi$. The interference of N is defined as the maximum number of transmission ranges in $B_{P_\pi}(N) \cup I_\pi \cup O_\pi$ that cover any given point of P_π. The transmission ranges in $O_\pi \cup I_\pi$ are given as pairs $(p, q) \in P^2$, where p is the center and q a point on the boundary of the range.

Each range in $O_\pi \cup I_\pi$ covers a boundary point of P_π. Since it is known that there is always an assignment with interference $O(\log n)$ (see [12] and Observation 5.1), no point of P lies in more than $O(\log n)$ ranges of $B_P(N^*)$. Thus, we can assume that $|I_\pi \cup O_\pi| = O(\log n)$, and the total number of subproblems is $n^{O(\log n)}$.

A subproblem π can be solved recursively as follows. Let A be the points in P_π to the left of r_π, and B the points in P_π to the right of r_π. We enumerate all pairs (σ, ρ) of subproblems with $P_\sigma = A$ and $P_\rho = B$, and we connect the roots r_σ and r_ρ to r_π. Then we check whether I_π, O_π, I_σ, O_σ, I_ρ, and O_ρ are *consistent*. This means that O_σ contains all ranges from O_π with center in A plus the range for the edge $r_\sigma r_\pi$ (if it does not lie in O_π yet). Furthermore, O_σ may contain additional ranges with center in A that cover points in $P_\pi \setminus A$ but not in $P \setminus P_\pi$. The set I_σ must contain all ranges in I_π and O_ρ that cover points in A, as well as the range from O_π with center r_π, if it exists and if it covers a point in A. The conditions for ρ are analogous.

Let N_π be the valid assignment for π obtained by taking optimal valid assignments N_σ and N_ρ for σ and ρ and by adding edges from r_σ and r_ρ to r_π. The interference of N_π is then defined with respect to the ranges in $B_{P_\pi}(N_\pi) \cup I_\pi$

plus the range with center r_π in O_π (the other ranges of O_π must lie in $B_{P_\pi}(N_\pi)$. We take the pair (σ, ρ) of subproblems which minimizes this interference. This step takes $n^{O(\log n)}$ time, because the number of subproblem pairs is $n^{O(\log n)}$ and the overhead per pair is polynomial in n.

The recursion ends if P_π contains a single point r_π. If O_π contains only one range, namely the edge from r_π to its parent, the interference of π is given by $|I_\pi| + 1$. If O_π is empty or contains more than one range, then the interference for π is ∞.

To find the overall optimum, we start the recursion with $P_\pi = P, O_\pi = I_\pi = \emptyset$ and every possible root, taking the minimum of all results. By implementing the recursion with dynamic programming, we obtain the following result.

Theorem 4.7. *Let $P \subset \mathbb{R}$ with $|P| = n$. The optimum interference of P can be found in time $n^{O(\log n)}$.* $\square$

Theorem 4.7 can be improved slightly. The number of subproblems depends on the maximum number of transmission ranges that cover the boundary points of P_π in an optimum assignment. This number is bounded by the optimum interference of P. Using exponential search, we get the following theorem.

Theorem 4.8. *Let $P \subset \mathbb{R}$ with $|P| = n$. The optimum interference OPT for P can be found in time $n^{O(\text{OPT})}$.* $\square$

5   Further Structural Properties in One Dimension

In this section, we explore further structural properties of optimal valid receiver assignments for one-dimensional point sets. It is well known that for any n-point set P, there always exists a valid assignment $\widetilde{N}$ with $I(\widetilde{N}) = O(\log n)$. Furthermore, there exist point sets such that any valid assignment N for them must have $I(N) = \Omega(\log n)$ [12]. For completeness, we include proofs for these facts in Sect. 5.1. In Sect. 5.2, Below we show that there may be an arbitrary number of left-right turns in an optimal solution. To the best of our knowledge, this result is new, and it shows that in a certain sense, Theorem 4.6 cannot be improved.

5.1   Nearest Neighbor Algorithm and Lower Bound

First, we prove that we can always obtain interference $O(\log n)$, a fact used in Sect. 4.2. This is achieved by the *Nearest-Neighbor-Algorithm* (NNA) [7,12]. It works as follows.

At each step, we maintain a partition $\mathcal{S} = \{S_1, S_2, \ldots, S_k\}$ of P, such that the convex hulls of the S_i are disjoint. Each set S_i has a designated sink $r_i \in S_i$ and an assignment $N : S_i \to S_i$ such that the graph $G_{S_i}(N_i)$ is acyclic and has r_i as the only sink. Initially, $\mathcal{S}$ consists of n singletons, one for each point in P. Each point in P is the sink of its set, and the assignments are trivial.

Now we describe how to go from a partition $\mathcal{S} = \{S_1, \ldots, S_k\}$ to a new partition $\mathcal{S}'$. For each sink $r_i \in S_i$, we define the successor $Q(r_i)$ as the closest point to r_i in $P \backslash S_i$. We will ensure that this closest point is unique in every round after the first. In the first round, we break ties arbitrarily Consider the directed graph R that has vertex set P and contains all edges from the component graphs $G_{S_i}(N_i)$ together with edges $r_i Q(r_i)$, for $i = 1, \ldots, k$. Let $S_1', S_2', \ldots, S_{k'}'$ be the components of R. Each such component S_j' contains exactly one cycle, and each such cycle contains exactly two sinks r_a and r_{a+1}. Pick $r_j' \in \{r_a, r_{a+1}\}$ such that the distances between r_j' and the closest points in the neighboring components S_{j-1}' and S_{j+1}' are distinct (if they exist). At least one of r_a and r_{a+1} has this property, because r_a and r_{a+1} are distinct. Suppose that $r_j' = r_a$ (the other case is analogous). We make r_a the new sink of S_j', and we let N_j' be the union of $r_{a+1}Q(r_{a+1})$ and the assignments N_i for all components $S_i \subseteq S_j$. Clearly, N_j' is a valid assignment for S_j'. We set $\mathcal{S}' = \{S_1', \ldots, S_{k'}'\}$. This process continues until a single component remains.

Observation 5.1. *The nearest neighbor algorithm ensures interference at most* $\lceil \log n \rceil + 2$.

Proof. Since each component in $\mathcal{S}$ is combined with at least one other component of $\mathcal{S}$, we have $k' \leq \lfloor k/2 \rfloor$, so there are at most $\lceil \log n \rceil$ rounds.

Now fix a point $p \in P$. We claim that in the interference of p increases by at most 1 in each round, except for possibly two rounds in which the interference increases by 2. Indeed, in the first round, the interference increases by at most 2, since each point connects to its nearest neighbor (the increase by 2 can happen if there is a point with two nearest neighbors). In the following rounds, if p lies in the interior of a connected component S_i, its interference increases by at most 1 (through the edge from r_i to $Q(r_i)$). If p lies on the boundary of S_i, its interference may increase by 2 (through the edge between r_i and $Q(r_i)$ and the edge that connects a neighboring component to p). In this case, however, p does not appear on the boundary of any future components, so the increase by 2 can happen at most once. □

Next, we show that interference $\Omega(\log n)$ is sometimes necessary. We make use of the points sets P_i constructed in Sect. 5.

Corollary 5.2. *For every n, there exists a point set Q_n with n points such that every valid assignment for N has interference $\lfloor \log n \rfloor$.*

Proof. Take the point set $P_{\lfloor \log n \rfloor}$ from Sect. 5 and add $n - 2^{\lfloor \log n \rfloor}$ points sufficiently far away. The bound on the interference follows from Proposition 5.3. □

5.2   Bends

In Theorem 4.6 we proved that there always exists an optimal solution with the BST-property. Now, we will show that the structure of an optimal solution cannot be much simpler than that. Let $P \subset \mathbb{R}$ be finite and let N be a valid receiver assignment for P. A *bend* in $G_P(N)$ is an edge between two non-adjacent points. We will show that for any k there is a point set Q_k such that any optimal assignment for Q_k has at least k bends.

For this, we inductively define sets P_0, P_1, ... as follows. For each P_i, let ℓ_i denote the diameter of P_i. P_0 is just the origin (and $\ell_0 = 0$). Given P_i, we let P_{i+1} consist of two copies of P_i, where the second copy is translated by $2\ell_i + 1$ to the right, see Fig. 4. By induction, it follows that $|P_i| = 2^i$ and $\ell_i = (3^i - 1)/2$.

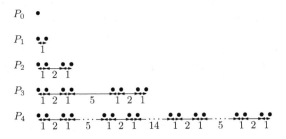

Fig. 4. Inductive construction of P_i.

Proposition 5.3. *Every valid assignment for P_i has interference at least i.*

Proof. The proof is by induction on i. For P_0 and P_1, the claim is clear.

Now consider a valid assignment N for P_i with sink r. Let Q and R be the two P_{i-1} subsets of P_i, and suppose without loss of generality that $r \in R$. Let E be the edges that cross from Q to R. Fix a point $p \in Q$, and let q be the last vertex on the path from p to r that lies in Q. We replace every edge $ab \in E$ with $a \neq q$ by the edge aq. By the definition of P_i, this does not increase the interference. We thus obtain a valid assignment $N' : Q \to Q$ with sink q such that $I(N) \geq I(N') + 1$, since the ball for the edge between q and R covers all of Q. By induction, we have $I(N') \geq i - 1$, so $I(N) \geq i$, as claimed. $\square$

Lemma 5.4. *For $i \geq 1$, there exists a valid assignment N_i for P_i that achieves interference i. Furthermore, N_i can be chosen with the following properties: (i) N_i has the BST-property; (ii) the leftmost or the rightmost point of P_i is the root of $G_{P_i}(N_i)$; (iii) the interference at the root is 1, the interference at the other extreme point of P_i is i.*

Proof. We construct N_i inductively. The point set P_1 has two points at distance 1, so any valid assignment has the claimed properties.

Given N_i, we construct N_{i+1}: recall that P_{i+1} consists of two copies of P_i at distance $\ell_i + 1$. Let L be the left and R the right copy. To get an assignment N_{i+1} with the leftmost point as root, we use the assignment N_i with the left point as root for L and for R, and we connect the root of R to the rightmost point of L. This yields a valid assignment. Since the distance between L and R is $\ell_i + 1$, the interference for all points in R increases by 1. The interferences for L do not change, except for the rightmost point, whose interference increases

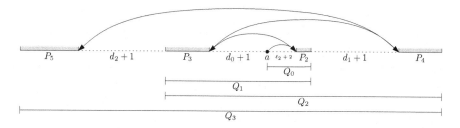

Fig. 5. The structure of Q_3. The arrows indicate the bends of an optimal assignment.

by 1. Since $|L| \geq 2$, the desired properties follow by induction. The assignment with the rightmost point as root is constructed symmetrically. □

The point set Q_k is constructed recursively. Q_0 consists of a single point $a = 0$ and a copy R_2 of P_2 translated to the right by $\ell_2 + 1$ units. Let d_{k-1} be the diameter of Q_{k-1}. To construct Q_k from Q_{k-1}, we add a copy R_{k+2} of P_{k+2}, at distance $d_{k-1} + 1$ from Q_k. If k is odd, we add R_{k+2} to the left, and if k is even, we add R_{k+2} to the right; see Fig. 5.

Theorem 5.5. *We have the following properties: (i) the diameter d_k is $(3^{k+3} - 2^{k+3} - 1)/2$; (ii) the optimum interference of Q_k is $k+2$; and (iii) every optimal assignment for Q_k has at least k bends.*

Proof. By construction, we have $d_0 = 9$ and $d_k = 2d_{k-1} + 1 + \ell_{k+2}$, for $k \geq 1$. Solving the recursion yields the claimed bound.

In order to prove (ii), we first exhibit an assignment N for Q_k that achieves interference $k + 2$. We construct N as follows: first, for $i = 2, \ldots, k + 1$, we take for R_i the assignment N_i from Lemma 5.4 whose root is the closest point of P_i to a. Then, we connect a to the closest point in R_2, and for $i = 2, \ldots, k + 1$, we connect the root of R_i to the root of R_{i+1}. Using the properties from Lemma 5.4, we can check that this assignment has interference $k + 2$.

Next, we show that all valid assignments for Q_k have interference at least $k + 2$. Let N be an assignment for Q_k. Let p be the leftmost point of R_{k+2}, and let q be the last point on the path from p to the root of N that lies in R_{k+2}. We change the assignment N such that all edges leaving R_{k+2} now go to q. This yields a valid assignment $\widetilde{N}$ for R_{k+2} with root q. Thus, $I(\widetilde{N}) \geq k + 2$, by Proposition 5.3. Hence, by construction, $I(N) \geq I(\widetilde{N}) \geq k + 2$, since $d_k \geq \ell_{k+2}$.

For (iii), let N be an optimal assignment for Q_k. We prove by induction that the root of N lies in R_{k+2}, and that N has k bends, all of which originate outside of R_{k+2}. As argued above, we have $I(N) = k + 2$. As before, let p be the leftmost point of R_{k+2} and q the last point on the path from p to the root of $G_{Q_k}(N)$. Suppose that q is not the root of N. Then q has an outgoing edge that increases the interference of all points in R_{k+2} by 1. Furthermore, by constructing a valid assignment $\widetilde{N}$ for R_{k+2} as in the previous paragraph, we see that the interference in N of all edges that originate from $P_{k+2} \setminus q$ is at least $k + 2$. If follows that $I(N) \geq k + 3$, although N is optimal.

Thus, the root r of N lies in R_{k+2}. Let b be a point outside R_{k+2} with $N(b) \in R_{k+2}$. The outgoing edge from b increases the interference of all points in $Q_k \setminus R_{k+2}$ by 1. Furthermore, we can construct a valid assignment $\widehat{N}$ for $Q_k \setminus R_{k+2}$ by redirecting all edges leaving Q_{k-1} to b. By construction, $I(\widehat{N}) \leq k+1$, so by (ii), $\widehat{N}$ is optimal for Q_{k-1} with interference $k+1$. By induction, $\widehat{N}$ has its root in R_{k+1} and has at least $k-1$ bends, all of which originate outside R_{k+1}. Thus, b must lie in R_{k+1}. Since b was arbitrary, it follows that all bends of $\widehat{N}$ are also bends of N. The edge from b in N is also a bend, so the claim follows.   □

6   Conclusion

We have shown that interference minimization in two-dimensional planar sensor networks is NP-complete. In one dimension, there exists an algorithm that runs in quasi-polynomial time, based on the fact that there are always optimal solutions with the BST-property. Since it is generally believed that NP-complete problems do not have quasi-polynomial algorithms, our result indicates that one-dimensional interference minimization is probably not NP-complete. However, no polynomial-time algorithm for the problem is known so far. Furthermore, our structural result in Sect. 5 indicates that optimal solutions can exhibit quite complicated behavior, so further ideas will be necessary for a better algorithm. In two dimensions naturally approximation algorithms (or approximation lower bounds.

Acknowledgments. We would like to thank Maike Buchin, Tobias Christ, Martin Jaggi, Matias Korman, Marek Sulovský, and Kevin Verbeek for fruitful discussions.

References

1. Bilò, D., Proietti, G.: On the complexity of minimizing interference in ad-hoc and sensor networks. Theor. Comput. Sci. **402**(1), 43–55 (2008)
2. Buchin, K.: Minimizing the maximum interference is hard (2008). arXiv:0802.2134
3. Burkhart, M., von Rickenbach, P., Wattenhofer, R., Zollinger, A.: Does topology control reduce interference? In: Proceedings of the 5th ACM International Symposium on Mobile Ad Hoc Networking and Computing (MobiHoc), pp. 9–19 (2004)
4. Fussen, M., Wattenhofer, R., Zollinger, A.: On interference reduction in sensor networks. Technical report 453, ETH Zürich, Department of Computer Science (2004)
5. Halldórsson, M.M., Tokuyama, T.: Minimizing interference of a wireless ad-hoc network in a plane. Theor. Comput. Sci. **402**(1), 29–42 (2008)
6. Johansson, T., Carr-Motyčková, L.: Reducing interference in ad hoc networks through topology control. In: Proceedings of the 2005 Joint Workshop on Foundations of Mobile Computing (DIALM-POMC), pp. 17–23 (2005)
7. Korman, M.: Minimizing interference in ad hoc networks with bounded communication radius. Inf. Process. Lett. **112**(19), 748–752 (2012)

8. Meyer auf der Heide, F., Schindelhauer, C., Volbert, K., Grünewald, M.: Congestion, dilation, and energy in radio networks. Theory Comput. Syst. **37**(3), 343–370 (2004)
9. Moscibroda, T., Wattenhofer, R.: Minimizing interference in ad hoc and sensor networks. In: Proceedings of the 2005 Joint Workshop on Foundations of Mobile Computing (DIALM-POMC), pp. 24–33 (2005)
10. Papadimitriou, C.H., Vazirani, U.V.: On two geometric problems related to the traveling salesman problem. J. Algorithms **5**(2), 231–246 (1984)
11. von Rickenbach, P., Schmid, S., Wattenhofer, R., Zollinger, A.: A robust interference model for wireless ad-hoc networks. In: Proceedings of the 19th IEEE International Parallel and Distributed Processing Symposium (IPDPS), p. 239 (2005)
12. von Rickenbach, P., Wattenhofer, R., Zollinger, A.: Algorithmic models of interference in wireless ad hoc and sensor networks. IEEE/ACM Trans. Netw. **17**(1), 172–185 (2009)
13. Tan, H., Lou, T., Wang, Y., Hua, Q.-S., Lau, F.: Exact algorithms to minimize interference in wireless sensor networks. Theor. Comput. Sci. **412**(50), 6913–6925 (2011)

Minimum Latency Aggregation Scheduling in Wireless Sensor Networks

Jonathan Gagnon and Lata Narayanan[(✉)]

Department of Computer Science and Software Engineering,
Concordia University, Montréal, QC, Canada
lata@cs.concordia.ca

Abstract. In wireless sensor networks, sensor nodes are used to collect data from the environment and send it to a data collection point or a *sink node* using a converge cast tree. Considerable savings in energy can be obtained by *aggregating* data at intermediate nodes along the way to the sink.

We study the problem of finding a minimum latency aggregation tree and transmission schedule in wireless sensor networks. This problem is referred to as Minimum Latency Aggregation Scheduling (MLAS) in the literature and has been proven to be NP-Complete even for unit disk graphs. For sensor networks deployed in a linear domain, that are represented as unit interval graphs, we give a 2-approximation algorithm for the problem. For k-regular unit interval graphs, we give an optimal algorithm: it is guaranteed to have a latency that is within one time slot of the optimal latency. We also give tight bounds for the latency of aggregation convergecast for grids and tori.

1 Introduction

A major application area for wireless sensor networks is to collect information about the environment in which they are deployed. In most applications, the collected information is sent to a selected node called the *sink*. This communication pattern is called *convergecast* [13] and has been studied extensively in the context of WSNs. Convergecasting is usually done by building a tree rooted at, and directed towards, the sink and by routing packets along the tree's edges toward the sink. Properly scheduling the nodes' transmissions is important to avoid possible interference.

Sensor nodes are powered by small batteries, and in many applications, it is infeasible or very expensive to replace or recharge the battery. Therefore, energy efficiency is an overriding concern in the design of communication protocols for wireless sensor networks. Since the radio is by far the most power-hungry element of a sensor node [5], any reduction in the transmitted data can be translated into energy savings for the sensor node. Even though the processing power of sensor nodes is limited, it is usually sufficient for simple computations. This allows for some processing of the raw data to be done before its transmission, and the cost of this local processing is negligible compared to the cost of transmissions.

© Springer-Verlag Berlin Heidelberg 2015
J. Gao et al. (Eds.): ALGOSENSORS 2014, LNCS 8847, pp. 152–168, 2015.
DOI: 10.1007/978-3-662-46018-4_10

For example, a sensor node could compress the sensor readings, or send a simple function of the sensor readings, thus reducing the size of the packets it sends. Additionally, in a convergecast operation, a sensor node could *combine* multiple packets received from its children, perhaps with its own data, before forwarding it to its parent in the tree to reduce the number of its own transmitted packets. Finally, in some applications, the information can be *aggregated* along the way to the sink, using a specific aggregation function. For example, if the sink node is interested in finding the maximum or average temperature in a region, each sensor node monitoring the region can easily aggregate the data received from its children with its own and simply send one packet containing the result. In large networks, this can dramatically reduce the total number and size of packets sent, because each node sends only one packet and the total number of packets sent is always equal to $n-1$. Without aggregation, $\Omega(n^2)$ packets are needed in the worst case.

In this paper, we study the problem of convergecast with aggregation in wireless sensor networks that are modeled by unit disk graphs: two nodes are connected if they are within each other's transmission range. Convergecast is performed along a spanning tree of the graph, and each tree link must be scheduled to transmit at a transmission slot so that (a) links that potentially interfere with each other are scheduled to transmit in different time slots, and (b) every node transmits only after all its children in the tree have transmitted, thereby ensuring aggregation can take place. We use the graph-based model of interference: two links (u_1, v_1) and (u_2, v_2) are said to interfere at the receiver v_1, if v_1 is within the transmission range of u_2. The *latency* of a convergecast operation is the time taken for the sink node to receive the data from all the nodes, which is defined as the time slot immediately after the largest transmission time of any node in a valid schedule. The problem of minimizing this latency is referred to as the Minimum Latency Aggregation Scheduling (MLAS) problem [19].

1.1   Related Work

Broadcast has been studied extensively in various models of communication for wired networks [6], as well as in wireless networks [4]. Convergecast is sometimes referred to as the gathering problem and the latency of convergecast *without* aggregation has been studied in [1,2,11,16,18]. As already mentioned, in this paper, we are studying convergecast with aggregation. In the wired setting, the time for broadcast on a tree is the same as the time for aggregation convergecast. However, in the wireless setting, these two times are usually quite different because of the broadcast nature of wireless transmissions. For example, in a tree where the root node has two children, broadcast can be accomplished in one time slot, while two time slots are needed for convergecast. Secondly, scheduling transmissions in a spanning tree of a graph is very different in wired versus wireless networks because of interference caused by *links that are in the underlying graph, but are not in the tree*. For example, Fig. 1 illustrates that in a clique of size 4, the latency of wired aggregation convergecast is 3 while the latency of wireless aggregation convergecast is 4. Aggregation convergecast can

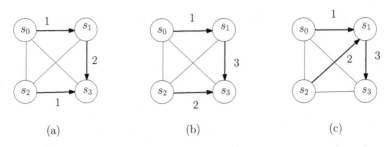

Fig. 1. Convergecast tree links are shown in bold, s_3 is the sink node, and transmission time slots are shown adjacent to tree links. (a) Wired aggregation convergecast can be scheduled with two time slots. (b) With the same tree, three time slots are needed in the wireless setting since the links (s_2, s_3) and (s_0, s_1) interfere. (c) A different convergecast tree and schedule, also requiring three time slots.

be scheduled in 2 time slots in a wired setting, as shown in Fig. 1(a), and thus has latency 3. However, when using the same tree as (a) in a wireless setting, we require three time slots as shown in Fig. 1(b). In fact, at most one link in the clique can be scheduled at any time, since any two links interfere with each other. Since there are three links in any spanning tree for the 4-clique, we need three time slots to schedule all transmissions for aggregation convergecast in the wireless setting. Figure 1(c) shows a different spanning tree for the same graph that also requires three transmission time slots. To summarize, convergecast in wireless networks is a different problem than either broadcast in wireless networks or convergecast in wired networks.

Data aggregation has been proposed early on in WSNs to reduce the energy usage of sensor nodes and improve the network lifetime. Krishnamachari et al. [12] demonstrated that significant energy savings could be achieved by using data aggregation. Routing protocols such as Directed Diffusion [9] and LEACH [7] incorporate data aggregation, and TAG and FILA [14,20] provide aggregation services when querying nodes in a sensor network.

In this paper, we consider the problem of minimizing the latency of aggregation convergecast, called the MLAS (Minimum Latency Aggregation Scheduling Problem) [19]. The problem has also been called aggregation convergecast [15] and MDAT (Minimum Data Aggregation Time) [3] in the literature. Chen et al. [3] proved that the MLAS problem is NP-complete, even for unit disk graphs. They also gave a centralized $(\Delta - 1)$-approximation algorithm named Shortest Data Aggregation (SDA), where Δ is the maximum degree of nodes in the graph. Huang et al. [8] designed a centralized algorithm based on Maximal Independent Sets (MIS) and with a latency bound of $23R + \Delta - 18$, where R is the maximum distance between the sink and any other node. Using a similar MIS approach, Wan et al. [19] proposed three new centralized algorithms, SAS, PAS and E-PAS, with latency bounds of $15R + \Delta - 4$, $2R + O(\log R) + \Delta$ and $(1 + O(\log R/\sqrt[3]{R}))R + \Delta$ respectively. However, there is no proven bound on the approximation ratio on the algorithms in [8,19]. Malhotra et al. [15] present two centralized algorithms, one is a tree construction algorithm called BSPT

(Balanced Shortest Path Tree) and the other is a ranking/priority-based scheduling algorithm called WIRES (Weighted Incremental Ranking for convergEcast with aggregation Scheduling), and showed that a combination of the two algorithms performed better in practice than previously proposed algorithms. Distributed algorithms for aggregation convergecast were studied in [10, 21, 22].

Note that for tree topologies, convergecast in a wireless network is the same as convergecast in a wired network, which in turn can be trivially derived from broadcast in a wired tree, and can be solved optimally [17]. To the best of our knowledge, no paper has addressed the MLAS problem for specific topologies like grids, tori and unit interval graphs. In our work, we present new lower bounds for all these topologies, and we present optimal algorithms for grids and tori as well as near-optimal algorithms for unit interval graphs and regular unit interval graphs.

1.2   Our Results

We study aggregation convergecast in unit interval graphs, which represent sensor networks deployed in a linear domain. We provide non-trivial lower bounds for the latency of aggregation convergecast in unit interval graphs, and give a 2-approximation algorithm. Our 2-approximation algorithm compares favorably to the best known approximation ratio of $\Delta - 1$ for general graphs. For k-regular unit interval graphs, we provide an algorithm which is guaranteed to have a latency that is within one time slot of the optimal latency, and is exactly optimal for many cases. We also prove lower bounds for the latency of aggregation convergecast for grids and tori, and we provide algorithms with matching upper bounds.

1.3   Model and Problem Statement

We assume that the nodes are synchronized and that they share the same wireless channel. All nodes are stationary and their transmission range is assumed to be constant and identical. The interference radius is assumed to be equal to the transmission range. Time is assumed to be slotted, and each node is scheduled to transmit in a given slot. Two nodes can transmit in the same time slot so long as their transmissions do not interfere. We assume that nodes can use an aggregation function that takes as input upto n data elements and produces a single element as output. Examples of such functions are *Min*, *Max*, *Sum* and *Count*.

Given a set of sensor nodes $S = \{s_0, s_1, \ldots, s_{n-1}\}$ with s_{n-1} being the sink node and where each node has a data item that it wants to send to the sink node, the problem we are interested in is to find a transmission schedule to send all the aggregated data to the sink in such a manner that each node transmits exactly once. Clearly, this implies that the schedule is interference-free.

More precisely, given a wireless network represented by a graph $G = (V, E)$, we define a *valid schedule* for G to be a spanning tree T of G rooted at and directed towards the sink node $s \in V$, and an assignment $A : V - \{s_{n-1}\} \to \mathbb{Z}^+$ of time slots to the nodes of the graph (except the sink) such that

1. $v \in children(u) \implies A(u) > A(v)$
2. $(u,v) \in T$ and $(w,v) \in G \implies A(u) \neq A(w)$

The first condition ensures that a node transmits after its children have all transmitted thereby guaranteeing aggregation of data, and the second condition ensures that transmissions are free of intereference. The *latency* of a valid schedule A for a graph G is denoted by $L(G, A)$, and is defined as $L(G, A) = max_{v \in V}\{A(v)\}+1$. The MLAS problem is now formally defined as follows: Given a graph $G = (V, E)$, find a valid schedule of minimum latency for G.

The following lemma is used extensively in our proofs:

Lemma 1. *For any graph $G = (V, E)$ with a sink node s, and a valid schedule A for G*

$$L(G, A) \geq max_{v \in V}\{A(v) + dist(v, s)\}$$

Proof. If node v transmits at time $A(v)$, its packet needs at least $dist(v, s)$ time steps to get to the sink.

We consider a unit interval graph $G = (V, E)$ of size n, where $V = \{s_0, ..., s_{n-1}\}$ and where s_{n-1} is the sink node. We assume that all sensor nodes are located at distinct locations. The nodes are sorted in descending order of distance from the sink which means s_0 is the farthest node from the sink. The node s_j is called a *forward* neighbor of s_i if it is closer to the sink than s_i (i.e. if $j > i$), otherwise it is called a *backward* neighbor. We denote by $dist(s_i, s_{n-1})$ the minimum graph distance between a node s_i and the sink. For any subset $V' \subseteq V$, we denote by $dist(V', s_{n-1})$ the minimum distance between any of the nodes in V' to the sink node.

We also study a special kind of unit interval graph where every node s_i is connected to $\{s_j \mid max\{0, i - k\} \leq j \leq min\{n - 1, i + k\}\}$. In other words, every node (except for the last k nodes) has k forward neighbors and every node (except for the first k nodes) has k backward neighbors. We call such a graph a *k-regular unit interval graph*.

2   Unit Interval Graphs

In this section, we give a lower bound for the latency of aggregation convergecast in unit interval graphs, and give a 2-approximation algorithm for the MLAS problem. Lemma 1 already gives a lower bound for the problem. To improve the lower bound for unit interval graphs, we consider a special set of cliques in the graph. We start with some elementary observations.

Lemma 2. *Given a unit interval graph G and clique C in G that contains s_0, no two nodes in C can transmit in the same time slot.*

Proof. Assume for the purpose of contradiction that s_i and s_j with $i < j$ are both part of a clique that contains s_0, and they both transmit in the same time slot G with s_i transmitting to node s_k. Clearly if s_k is a forward neighbor of s_i, it is also a neighbor of s_j in the unit interval graph since $j > i$. Also if s_k is a backward neighbor of s_i, it follows that $0 \leq k < i$, and therefore s_k is in the clique C and is a neighbor of s_j. Thus the transmission of s_j (to any recipient) would interfere with the transmisison of s_i to s_k, contradicting the validity of the schedule.

For any clique that does not contain the first node s_0, there is in fact a possibility of two nodes being able to transmit in the same time slot, as shown in the lemma below:

Lemma 3. *Given a unit interval graph G, and s_i and s_j with $i < j$, let C be a clique containing both s_i and s_j. At most two nodes in C can transmit in the same time slot. Furthermore if the transmissions (s_i, s_k) and (s_j, s_ℓ) are scheduled in the same time slot, then $k < i$, $j < \ell$, and $s_k, s_\ell \notin C$.*

Proof. If $k > i$, or if $s_k \in C$, then s_j's transmission will interfere with s_k's reception. If $\ell < j$, or if $s_\ell \in C$, then s_i's transmission will interfere with s_ℓ's reception.

We denote by $last(s_i)$ the last neighbor of s_i, i.e. the neighbor of s_i that is the closest to the sink. We partition the nodes of the graph into disjoint cliques $C_0, ..., C_m$ as follows:

$$C_i = \{s_{j_i}, \ldots, s_k\} \text{ with } k = j_{i+1} - 1$$

where $s_{j_0} = s_0$ and $s_{j_i} = last(s_{j_{i-1}})$ for $0 < j \leq m$. Thus, the first node of the clique C_i is the last neighbor of the first node of the previous clique C_{i-1}. We denote $|C_i|$ by a_i. Figure 2 shows an example of a graph with four cliques. Observe that the last clique $C_m = \{s_{n-1}\}$ and $s_{j_m} = s_{n-1}$.

The following lower bound on the latency of convergecast is straightforward:

Theorem 1. *In a unit interval graph of size n, divided into m cliques as described above, any MLAS scheduling algorithm must have latency at least*

$$max_{\{1 \leq i \leq m-1\}} \left\lceil \frac{a_{i-1} + a_i}{2} \right\rceil + dist(C_i, s_{n-1})$$

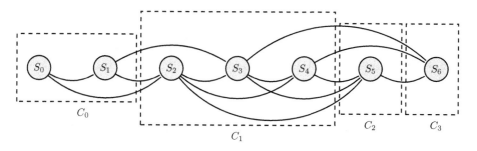

Fig. 2. Illustration of the division of a graph with 3 cliques where $C_0 = \{s_0, s_1\}$, $C_1 = \{s_2, s_3, s_4\}$, $C_2 = \{s_5\}$, and $C_3 = \{s_6\}$.

Proof. Consider 2 consecutive cliques C_{i-1} and C_i with $0 < i \leq m-1$. We claim that at most two nodes from $C_{i-1} \cup C_i$ can transmit at the same time. Suppose for the purpose of contradiction that three nodes from this set transmit at time t. From Lemma 3, at most two of them can be from the same clique. So either two of the three claimed senders are from C_{i-1} and the remaining one in C_i, or two of them are in the clique C_i and the third sender in C_{i-1}. If two of the senders are from C_{i-1}, from Lemma 3, one of these nodes has to transmit to a forward neighbor outside C_{i-1}, that is, to a node $s \in C_i \cup \{s_{j_{i+1}}\}$. But this implies that no node in C_i can transmit at time t, since its transmission would interfere with the reception at node s, a contradiction. Similarly, if the two senders are in C_i, one of these nodes has to transmit to a node in C_{i-1}. Therefore, no node in C_{i-1} can transmit at time t, a contradiction. This shows that no more than 2 nodes in $C_{i-1} \cup C_i$ can transmit at the same time. The last node from $C_{i-1} \cup C_i$ cannot transmit before time $\lceil \frac{a_{i-1}+a_i}{2} \rceil$ and thus from Lemma 1, we have the desired lower bound for the latency of any MLAS scheduling algorithm.

We now present an algorithm that has latency as most twice as the above lower bound. We call it the *Hub Algorithm* and an example of the schedule produced by the algorithm is given in Fig. 3. The Hub Algorithm uses the cliques $C_0, ..., C_m$ defined earlier. Essentially, the first node in each clique is used as the aggregator or the *hub* for the nodes in the previous clique. This defines the convergecast tree: every node in clique C_{i-1} sends to s_{j_i}, the first node of C_i. Next, the transmission schedule, denoted by A_H, is built by a simple greedy approach. Starting with node s_0, we go through the nodes in order. For the first (hub) node in C_i, we assign it a time slot $1 + max_{v \in C_{i-1}}\{A(v)\}$. To every other (non-hub) node in C_i, we assign the lowest time slot not already in use by a node in the previous and current clique.

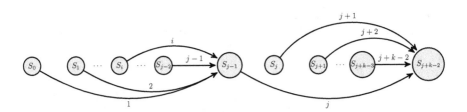

Fig. 3. In the Hub algorithm, intermediate nodes are selected and used as aggregators along the way to the sink. In this figure, $j = |C_0|$ and $k = |C_1|$. Note that the total number of time slots used by these 2 cliques is $j + k - 1$, assuming that s_{j+k-2} is not the sink and is scheduled at time $j + k - 1$.

Theorem 2. *The Hub Algorithm is a 2-approximation algorithm and builds a valid schedule in $\mathcal{O}(|V|)$ time.*

Proof. First we argue that the schedule A_H is interference-free. Observe that the only nodes that receive packets are the hubs, the first nodes in each clique.

Consider such a hub, say s_{j_i}. From the definition of the cliques, its neighbours are all in $C_{i-1} \cup C_i \cup s_{j_{i+1}}$. We claim that all of its neighbors are assigned different time slots. By design, $A_H(s_{j_{i+1}}) > A_H(s_{j_i}) > A_H(s)$ for all $s \in C_{i-1}$. Further, all non-hub nodes in C_{i-1} are assigned a slot different from previously assigned nodes in the previous two cliques, and the same is the case for non-hub nodes in C_i. Thus, the schedule is interference-free. Second, by design, the transmission slot of the hub nodes (aggregators) is after the transmission slot of its children in the convergecast tree. Therefore the schedule is a valid schedule.

Next, we determine an upper bound on the total number of time slots used. Let C_{i-2} and C_{i-1} be 2 consecutive cliques, the number of time slots needed by their nodes is $a_{i-2} + a_{i-1}$, or the total number of nodes in those 2 cliques. Therefore, each non-hub node in a clique C_{i-1} is assigned a time slot in the set $\{1, \ldots a_{i-2} + a_{i-1}\}$. From the defintion of $A_H(s_{j_i})$, it is easy to see that $A_H(s_{j_i}) = max\{A_H(s_{j_{i-1}}), a_{i-2} + a_{i-1}\} + 1$.

We now prove inductively that

$$A_H(s_{j_i}) = max_{2 \le \ell \le i}\{a_{\ell-2} + a_{\ell-1} + i - \ell + 1\}$$

for $2 \le i \le m$. It is easy to verify that node $A_H(s_{j_2}) = a_1 + a_2 + 1$. Supposing the claim to be true for $A_H(s_{j_{i-1}})$, observe that

$$A_H(s_{j_i}) = max\{A_H(s_{j_{i-1}}), a_{i-2} + a_{i-1}\} + 1 \tag{1}$$
$$= max\{max_{2 \le \ell \le i-1}\{a_{\ell-2} + a_{\ell-1} + i - 1 - \ell + 1\}, a_{i-2} + a_{i-1}\} + 1 \tag{2}$$
$$= max_{2 \le \ell \le i}\{a_{\ell-2} + a_{\ell-1} + i - \ell + 1\} \tag{3}$$

as needed. Recall that $s_{j_m} = s_{n-1}$, that is the first node (and hub) of the last clique is the last node s_{n-1}. Though s_{n-1} being the sink does not actually transmit, the time it has received data from its children is (with a slight abuse of notation) given by $A_H(s_{j_m})$; clearly $L(G, A_H) = A_H(s_{j_m})$. Therefore,

$$L(G, A_H) = A_H(s_{j_m}) = max_{2 \le \ell \le m}\{a_{\ell-2} + a_{\ell-1} + m - \ell + 1\}$$
$$= max_{2 \le \ell \le m}\{a_{\ell-2} + a_{\ell-1} + dist(C_{\ell-1}, s_{n-1})\}$$
$$\le 2max_{1 \le \ell \le m-1}\{\left\lceil \frac{a_{\ell-1} + a_\ell}{2} \right\rceil + dist(C_\ell, s_{n-1}\}\}$$

Combined with Theorem 1, this proves that the performance ratio of the Hub Algorithm is at most 2.

Assuming that the nodes are already sorted in order of index, the cliques can be found in $\mathcal{O}(|V|)$ time in the worst case. Every node is assigned as parent the last node in its clique which can be done in constant time. By maintaining separate lists for the set of time slots used in the previous two cliques and a set of available time slots, the schedule can be built in $\mathcal{O}(|V|)$ time.

There are instances where the Hub Algorithm uses $2k$ time slots while the optimal algorithm uses $k + 1$ slots, as shown in Figs. 5 and 6, thus the bound on the performance ratio is tight.

3   Optimal Algorithm for Regular Unit Interval Graphs

In this section, we consider convergecast algorithms for k-regular unit interval graphs; each node has k forward neighbors (except the last k nodes) and k backward neighbors (except the first k nodes). We prove an improved lower bound for such graphs, and then give an algorithm that matches this bound. We begin with a simple observation about distances in a k-regular unit interval graph.

Lemma 4. *In a k-regular unit interval graph, the number of hops between a node s_i and the sink s_{n-1} is given by*

$$dist(s_i, s_{n-1}) = \left\lceil \frac{n-1-i}{k} \right\rceil$$

We use the above lemma to obtain the following straightforward lower bound:

Theorem 3. *Let A be an algorithm for aggregation convergecast in a k-regular unit interval graph G with $n \geq k+1$ nodes. Then $L(G, A) \geq \left\lceil \frac{n-1}{k} \right\rceil + k$.*

Proof. The first $k+1$ nodes form a clique, and it follows from Lemma 2 that only one node from this clique can transmit at a time. Therefore, $k+1$ time slots are needed for these nodes to transmit their data. Suppose that $s_i \in \{s_0, \ldots, s_k\}$ is the last node in that group to transmit, the earliest it can transmit is at time $k+1$, and it follows from Lemmas 1 and 4 that for any valid schedule A:

$$L(G, A) \geq dist(s_i, s_{n-1}) + k + 1$$
$$= \left\lceil \frac{n-1-i}{k} \right\rceil + k + 1$$
$$\geq \left\lceil \frac{n-1-k}{k} \right\rceil + k + 1$$
$$= \left\lceil \frac{n-1}{k} \right\rceil + k$$

We now show that the above lower bound can be strengthened for large enough values of n and k. Let $n \geq 2k + 3$, $k > 2$ and consider the first $2k + 3$ nodes. We consider the following three disjoint groups of nodes: $G_0 = \{s_0, \ldots, s_k\}$, $G_1 = \{s_{k+1}, \ldots, s_{2k+1}\}$ and $G_2 = \{s_{2k+2}\}$. We first show a lemma based on distance. Let $A(s_i)$ be the transmission time of node s_i.

Lemma 5. *Suppose G is a k-regular graph with n vertices, with $n - 1 = ik + r$, where $0 \leq r < k$ and $r \notin \{1, 2\}$. For any aggregation convergecast schedule A, if $L(G, A) \leq \lceil \frac{n-1}{k} \rceil + k$, then:*

1. $\{A(s_i) \mid s_i \in G_0\} = \{1, 2, \ldots k + 1\}$
2. $\{A(s_i) \mid s_i \in G_1\} = \{1, 2, \ldots, k + 2\} - \{k + 1\}$

3. $A(s_{k+1}) \leq k$.
4. $A(s_{2k+2}) \leq k+1$.

Proof. Suppose there exists $s_i \in G_0$ such that $A(s_i) > k+1$. Then it follows from Lemma 1 that

$$L(G, A) \geq dist(s_i, s_{n-1}) + A(s_i)$$
$$\geq \left\lceil \frac{n-1-k}{k} \right\rceil + (k+2)$$
$$= \left\lceil \frac{n-1}{k} \right\rceil + k + 1$$

a contradiction to the assumption $L(G) \leq \lceil \frac{n-1}{k} \rceil + k$. Part (1) is now a consequence of the fact that $|G_0| = k+1$.

We can use an identical argument to show that $A(s_i) \leq k+2$ for any $s_i \in G_1$. However from part (1), there is a node in G_0 that transmits at time $k+1$, since there are $k+1$ nodes in G_0. This node is the last node to transmit in G_0 and must therefore transmit to a forward neighbor, that is, a node in G_1. It follows that $A(s_i) \neq k+1$ for any node in G_1. This establishes part (2) since $|G_1| = k+1$.

Next, suppose $s_{k+1} \in G_2$ transmits at time $\geq k+2$. Then

$$L(G, A) \geq dist(s_{k+1}, s_{n-1}) + A(s_i)$$
$$\geq \left\lceil \frac{n-1-(k+1)}{k} \right\rceil + k + 2$$
$$= \left\lceil \frac{ik+r-k-1}{k} \right\rceil + k + 2$$
$$\geq \left\lceil \frac{r-1}{k} \right\rceil + i - 1 + k + 2$$
$$= \left\lceil \frac{r}{k} \right\rceil + i + k + 1$$
$$= \left\lceil \frac{ik+r}{k} \right\rceil + k + 1$$
$$= \left\lceil \frac{n-1}{k} \right\rceil + k + 1$$

where $\lceil (r-1)/k \rceil = \lceil r/k \rceil$ because $r \neq 1$. Thus $A(s_{k+1}) \leq k+1$. But since $s_{k+1} \in G_1$, we have $A(s_{k+1}) \neq k+1$ (from part (2)). Therefore $A(s_{k+1}) \leq k$.

Using Lemma 1 and the fact that $r \notin \{1, 2\}$, we can show similarly that $A(s_{2k+2}) \leq k+2$. Finally, we use the fact that the last node to transmit in G_1 must transmit forward at time $k+2$ to see that s_{2k+2} cannot transmit it at time $k+2$, thus proving part (4).

Theorem 4. *Let A be an algorithm for aggregation convergecast, and G be a k-regular unit interval graph of size n with $n \geq 2k+3$, $k > 2$ and $(n-1) \bmod k \notin \{1, 2\}$. Then $L(G, A) \geq \left\lceil \frac{n-1}{k} \right\rceil + k + 1$.*

Proof. Suppose for the purpose of contradiction that there is an algorithm for aggregation convergecast with latency $\leq \lceil \frac{n-1}{k} \rceil + k$ time slots. From Lemma 5, the nodes in $\{s_0, s_1, \ldots, s_{k+1}\}$ must all use the time slots $\{1, \ldots, k+1\}$. That is, there is a node $s_i \in G_0$ that must transmit at the same time as s_{k+1}, say $A(s_{k+1}) = A(s_i) = t$. Since the only receiver that would not be affected by s_{k+1}'s transmission is s_0, it follows that s_i transmits to s_0 at time t where $1 \leq t \leq k-1$.

From Lemma 5, we know that $A(s_{2k+2}) \leq k+1$. Suppose $A(s_{2k+2}) \leq k$. Then s_{2k+2} transmits at the same time as a node $s_j \in G_1$, say at time $1 \leq t' \leq k$. Thus, s_j must transmit to the only non-neighbor of s_{2k+2} in G_1, that is, s_{k+1}. Since s_{k+1} can only transmit after it receives, it must be that $t > t'$. Furthermore, some node in G_0 must transmit at time t'. This node can only be s_0 as any other node's transmitting at time t' would interfere with s_{k+1}'s reception at time t'. Thus $A(s_0) = t'$. But s_0 receives at time t, and since s_0 can transmit only after it receives, it must be that $t' > t$, a contradiction.

Finally, suppose $A(s_{2k+2}) = k+1$. Then since there is a node s_ℓ in G_0 that transmits at the same time to a node in G_1, it must be that s_ℓ transmits to s_{k+1} as any other recipient in G_1 would experience interference with s_{2k+2}'s transmission. But this implies that $A(s_{k+1}) \geq k+1$, a contradiction to Lemma 5. This completes the proof.

We now proceed to give an algorithm that meets the bound of Theorem 4. We use the same partitioning of nodes into cliques as in Sect. 2. For k-regular graphs, observe that if m is the number of cliques:

$$C_i = \{s_{ik}, \ldots, s_{(i+1)k-1}\} \text{ for } 0 \leq i < m - 2$$
$$C_{m-1} = \{s_{(m-1)k}, \ldots, s_{n-1} - 1\} \text{ and finally}$$
$$C_m = \{s_{n-1}\}.$$

Clearly since s_{n-1} is the sole element of the last clique, we have $m = \lceil \frac{n-1}{k} \rceil + 1$.

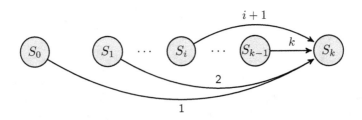

Fig. 4. Optimal solution using the Hub Algorithm when $n = k + 1$.

We start by building some intuition for our algorithm. Notice that if $n = k+1$, the schedule produced by the Hub Algorithm is optimal as shown in Fig. 4. However, the Hub Algorithm doesn't give an optimal solution for larger networks. For instance if $n = 2k+1$, the schedule provided by the Hub Algorithm as shown

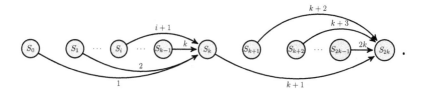

Fig. 5. Illustration of the tree and schedule built by the Hub Algorithm for a Regular Unit Interval Graph of size $2k + 1$. The schedule built with the Hub Algorithm has latency $2k + 1$ whereas an optimal solution for the same graph has latency $k + 2$.

in Fig. 5 is sub-optimal. In this specific example, since $\{s_0, s_1, s_k\}$ form a clique, by Lemma 2, they need at least $k+1$ time slots to transmit, giving a lower bound of latency $k+2$ for the latency. However, since nodes in C_0 transmit to s_k, nodes in C_1 cannot reuse the same time slots without causing interference at s_k. Thus, a total of $2k$ time slots are used by the Hub Algorithm for this example.

The proof of the lower bound of Theorem 1 suggests that in an optimal algorithm some nodes would transmit backward. Indeed, a better approach for the graph with $n = 2k + 1$ is to use node s_0 as a data aggregator for the nodes in C_0. Because s_0 is more than k nodes away from the nodes in C_1, nodes in C_1 are far enough to allow the same time slots to be reused without interference. However, when $n > 2k+1$, nodes in C_1 have to avoid interference with nodes in C_2 as well as avoiding interference with nodes in C_0. Nodes in C_1 cannot transmit backward to avoid interference with C_2, because there would be interference with a node in C_0. In order to minimize the chance of interference, a simple solution is to schedule nodes in C_1 to transmit to the closest forward neighbor. as shown in Fig. 6.

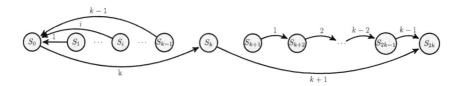

Fig. 6. Schedule for C_0 and C_1.

In general then, while assigning time slots to nodes in a clique, we aim to reuse as many time slots as possible while avoiding interference with their neighboring groups. We proceed to present our algorithm for constructing the tree and scheduling the nodes of a k-regular unit interval graph of size $n \geq 2k + 3$. As shown in Fig. 6, s_0 is used as a data aggregator for nodes in C_0 and will transmit its aggregated data to s_k at time k. We now describe the tree and schedule for the nodes in the other cliques. For $1 \leq i \leq m - 2$, let C_i be divided into 4 sub-groups A_i, B_i, E_i and D_i, and let $\alpha_i = min\{i, k-1\}$ and $\beta_i = min\{2i-2, k-1\}$.

The sub-groups are defined as follows:

$$A_i = \{s_{ik}\}$$

$$B_i = \begin{cases} \emptyset & \text{if } i = 1 \\ \{s_j \mid ik + 1 \le j \le ik + \alpha_i\} & \text{otherwise} \end{cases}$$

$$E_i = \{s_j \mid ik + \alpha_i + 1 \le j \le ik + \beta_i\}$$

$$D_i = \{s_j \mid ik + \beta_i + 1 \le j \le ik + k - 1\}$$

The following observations are straightforward:

Lemma 6. *1. $|A_i| = 1$ for $1 \le i \le m - 1$*
2. $|B_i| = \alpha_i$ for $1 \le i \le m - 1$. Furthermore, $B_2 = 2$, and $|B_i| = |B_{i-1}| + 1$ for $3 \le i \le k - 1$

3. $|E_i| = \begin{cases} \beta_i - \alpha_i & \text{if } i < k - 1 \\ 0 & \text{otherwise} \end{cases}$
Furthermore, $|E_3| = 1$ and $|E_i| = |E_{i-1}| + 1$ for $4 \le i \le k - 1$.
4. $|D_i| = \begin{cases} k - 1 - \beta_i & \text{if } i \le \frac{k}{2} \\ 0 & \text{otherwise} \end{cases}$
Furthermore $|D_1| = k - 1$ and $|D_i| = |D_{i-1}| - 2$ for $2 \le i \le (k-1)/2$.
5. $|B_{i-1}| \ge |E_i|$ for $i \ge 1$.

Proof. We only provide the proof of the last part here. If $i \ge k - 1$, then $|E_i| = 0$ so $|B_{i-1}| \ge |E_i|$. Otherwise $i < k - 1$, and $|B_{i-1}| = min\{i - 1, k - 1\} = i - 1$. However, $|E_i| = min\{2i - 2, k - 1\} - min\{i, k - 1\} = min\{2i - 2, k - 1\} - i = min\{i - 2, k - 1 - i\} \le i - 2$. Therefore $|B_{i-1}| \ge |E_i|$.

We are now ready to present our algorithm for k-regular unit interval graphs. For each node in each group, we specify the time slot it will transmit as well as the recipient. Figures 6, 7 and 8 (the last two in the appendix) together illustrate the

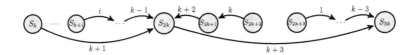

Fig. 7. Tree and schedule for C_1 and C_2. Nodes in C_1 transmit to their closest forward neighbor. This allows for $k - 3$ time slots to be reusable in C_2.

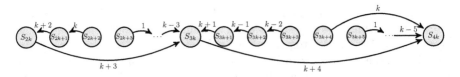

Fig. 8. Tree and schedule for C_2 and C_3. Some nodes in C_2 and C_3 will transmit to their closest backward neighbor to allow for more time slots to be reused in the next cliques.

schedule for a k regular interval graph of $4k+1$ nodes. Note that in the example, the time slot $k+1$ is assigned to node s_k while the schedule below specifies time slot $k+2$ for this node; both assignments work.

Algorithm k-regular Convergecast

1. $s_{ik} \in A_i$ transmits at time $k+i+1$. If $ik+k < n$ then the receiver is node s_{ik+k}, otherwise the receiver is node s_{n-1}.
2. We assign the time slots in $TB_i = \{1, \ldots, min(k+i, 2k-1)\} - T_{i-1}$ to the nodes in B_i, where T_{i-1} is the set of time slots assigned to nodes in C_{i-1}. Observe that $|T_i| = k$ and $|TB_i| \geq min\{k+i, 2k-1\} - k = min\{i, k-1\} = |B_i|$. The time slots are assigned in ascending order from the last node to the first node in B_i: node $s_j \in B_i$ transmits to node s_{j-1}.
3. For each $s_j \in E_i$, s_j transmits at the same time as node $s_{j-k-\alpha} \in B_{i-1}$. If $ik+k < n$, the receiver is node s_{ik+k}, otherwise the receiver is node s_{n-1}.
4. For each $s_j \in D_i$, s_j transmits to node s_{j+1} at time $j - (ik+\beta)$. Nodes in D_i will therefore use time slots in the range $1, \ldots, |D_i|$.

Theorem 5. *Algorithm k-regular Convergecast produces a valid interference-free schedule and has latency* $\left\lceil \dfrac{n-1}{k} \right\rceil + k + 1$. *This schedule can be built in* $\mathcal{O}(|V|)$ *time.*

Proof. It is easy to verify (see Fig. 6) that the schedule of the first $2k$ nodes is interference-free. We will show by induction that the schedule of the other nodes is also interference-free. Assume that nodes in the first i cliques have an interference-free schedule. It is clear from the description of the algorithm that each node in each subgroup in C_i is assigned a different time slot, so there cannot be any interference between them. We need to show that their transmissions never interfere with the transmission of a node in C_{i-1} or with previous subgroups in C_i itself.

The transmission of node s_{ik} is obviously interference-free since the latest time slot used in C_{i-1} is time $k+i$ and s_{ik} transmits at time $k+i+1$. Observe also that s_{ik} transmits after any other node in C_i and cannot interfere with any other node in C_i.

Nodes in B_i are assigned time slots from the set TB_i, which is disjoint from T_{i-1} by definition. Secondly, the largest time slot in TB_i is at most $i+k$, while the node in A_i has time slot $k+i+1$. Therefore, the transmission of a node in B_i does not interfere with any of the nodes in $C_{i-1} \cup A_i$.

Nodes in E_i are assigned the time slots of nodes in B_{i-1}, and by induction, cannot interfere with any node in $C_{i-1} \cup A_i \cup B_i - B_{i-1}$. Therefore, the only possibility of interference for nodes in E_i is with nodes in B_{i-1}. However, nodes in E_i transmit forward while nodes in B_{i-1} transmit backward. Furthermore, the distance between $s_j \in E_i$ that transmits at the same time as node $s_{j-k-\alpha} \in B_{i-1}$ is $k + \alpha > k$, therefore, their transmissions do not interfere.

Finally, we need to prove that the schedule of nodes in D_i is interference-free. Observe that time slots used by D_i are a strict subset of time slots used by D_{i-1}.

By induction, these are disjoint from all nodes in $C_{i-1} \cup A_i \cup B_i \cup E_I - D_{i-1}$. Therefore, nodes in D_i can only interfere with nodes in D_{i-1}. Suppose that $s_j \in D_i$ transmits to s_{j+1} and that $s_{j'} \in D_{i-1}$ transmits to $s_{j'+1}$, both at time $j - (ik + \beta)$. We need to verify that the distance between s_j and $s_{j'+1}$ as well as the distance between $s_{j'}$ and s_{j+1} are both greater than k. Based on the definition of our algorithm, we find that $j' = j - k - 2$. Therefore, the distance between s_j and $s_{ij'+1}$ is $j - (j - k - 1) = k + 1$ and that the distance between $s_{ij'}$ and s_{ij+1} is $(j + 1) - (j - k - 2) = k + 3$. Both distances are greater than k and therefore the transmissions do not interfere.

Thus the schedule given by Algorithm k-regular Convergecast is a valid schedule. We proceed to prove a bound on its latency. By the definition of the algorithm, the highest time slot used by the clique C_i is $k + i + 1$, so the last time slot will be used by the second-last clique. There are $\lceil \frac{n-1}{k} \rceil + 1$ cliques, numbered from 0 to $\lceil \frac{n-1}{k} \rceil$, and the second last clique is numbered $\lceil \frac{n-1}{k} \rceil - 1$. Thus, the last time slot used by the second-last group will be $k + (\lceil \frac{n-1}{k} \rceil - 1) + 1 = \lceil \frac{n-1}{k} \rceil + k$. Thus the latency of the algorithm is $\lceil \frac{n-1}{k} \rceil + k + 1$ as claimed.

As with the Hub Algorithm for unit interval graphs, we start by dividing the graph into cliques which can be done in $\mathcal{O}(|V|)$ time. The division of each clique into subgroups can be computed in constant time. It is straightforward to see from the description of the schedule that the remaining computations can be completed in $\mathcal{O}(|V|)$ time.

It follows from the lower bound in Theorem 4 that Algorithm k-regular Convergecast is optimal when $n \geq 2k + 3$, $k > 2$ and $(n - 1) \bmod k \notin \{1, 2\}$. For other values of n, the algorithm has latency at most one plus the optimal latency. Additionallly, if $n \leq 2k + 2$ or if $k = 2$, it is not hard to find a schedule with latency $(n - 1)/k + k$ that is optimal since it matches the lower bound of Theorem 3. The optimal latency for $n \geq 2k + 3$, $k > 2$, when $n - 1 \bmod k = 1$ or 2 remains open.

4   Optimal Convergecast for Grids and Tori

Due to lack of space, we simply present the results in this section; the proofs will appear in the full version (Fig. 9).

Theorem 6. *For any algorithm A for the MLAS problem on a grid G of k dimensions,*
$$L(G, A) \geq max\{dist(v_i, s) : i = 1, 2, \ldots, n\} + \sum_{j=1}^{k} C_j, \text{ where:}$$
$$C_j = \begin{cases} 1 & \text{if } G.size_j > 1 \text{ and } sink.pos_j == (G.size_j + 1)/2 \\ 0 & \text{otherwise} \end{cases}$$

Theorem 7. *For any algorithm A for the MLAS problem on a torus G of k dimensions,*
$$L(G, A) \geq \sum_{i=1}^{k} \lfloor \frac{G.size_i}{2} \rfloor + \sum_{j=1}^{k} C_j \text{ where:}$$
$$C_j = \begin{cases} 1 & \text{if } G.size_j > 1 \text{ and } G.size_j \text{ is odd} \\ 0 & \text{otherwise} \end{cases}$$

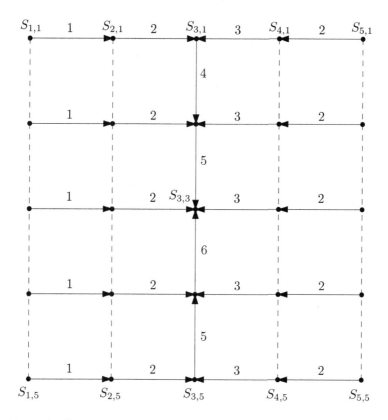

Fig. 9. Example of an optimal schedule for a 5×5 grid with the sink node located at $(3,3)$.

Theorem 8. *Given a grid or torus network of k dimensions, there is an algorithm that solves the MLAS problem optimally in $\mathcal{O}(|V|)$ time.*

References

1. Bermond, J.-C., Nisse, N., Reyes, P., Herve, R.: Minimum delay data gathering in radio networks. In: Proceedings of Ad Hoc-Now Conference (2009)
2. Bermond, J.-C., Peters, J.G.: Optimal gathering in radio grids with interference. Theor. Comput. Sci. **457**, 10–26 (2012)
3. Chen, X., Hu, X., Zhu, J.: Minimum data aggregation time problem in wireless sensor networks. In: Jia, X., Wu, J., He, Y. (eds.) MSN 2005. LNCS, vol. 3794, pp. 133–142. Springer, Heidelberg (2005)
4. Chlamtac, I., Kutten, S.: On broadcasting in radio networks-problem analysis and protocol design. IEEE Trans. Commun. **33**(12), 1240–1246 (1985)
5. Demirkol, I., Ersoy, C., Alagoz, F.: MAC protocols for wireless sensor networks: a survey. IEEE Commun. Mag. **44**(4), 115–121 (2006)
6. Hedetniemi, S.M., Hedetniemi, S.T., Liestman, A.L.: A survey of gossiping and broadcasting in communication networks. Networks **18**(4), 319–349 (1988)

7. Heinzelman, W.R., Chandrakasan, A., Balakrishnan, H.: Energy-efficient communication protocol for wireless microsensor networks. In: Proceedings of the 33rd Annual Hawaii International Conference, p. 10. IEEE (2000)
8. Huang, S.C.-H., Wan, P.-J., Vu, C.T., Li, Y., Yao, F.: Nearly constant approximation for data aggregation scheduling in wireless sensor networks. In: Proceedings of the 26th IEEE International Conference on Computer Communications (INFOCOM), pp. 366–372. IEEE (2007)
9. Intanagonwiwat, C., Govindan, R., Estrin, D.: Directed diffusion: a scalable and robust communication paradigm for sensor networks. In: Proceedings of the 6th Annual International Conference on Mobile Computing and Networking, pp. 56–67. ACM (2000)
10. Kesselman, A., Kowalski, D.R.: Fast distributed algorithm for convergecast in ad hoc geometric radio networks. J. Parallel Distrib. Comput. **66**(4), 578–585 (2006)
11. Klasing, R., Korteweg, L., Stougie, P., Marchetti-Spaccamela, A.: Data gathering in wireless networks. In: Koster, A., Muñoz, X. (eds.) Graphs and Algorithms in Communication Networks. Texts in Theoretical Computer Science: An EATCS Series, pp. 357–377. Springer, Heidelberg (2009)
12. Krishnamachari, B., Estrin, D., Wicker, S.: The impact of data aggregation in wireless sensor networks. In: Proceedings of the 22nd International Conference on Distributed Computing Systems, pp. 575–578. IEEE (2002)
13. Kulkarni, S.: TDMA service for sensor networks. In: Proceedings of the 24th International Conference on Distributed Computing Systems, pp. 604–609 (2004)
14. Madden, S., Franklin, M.J., Hellerstein, J.M., Hong, W.: TAG: a tiny aggregation service for ad-hoc sensor networks. ACM SIGOPS Oper. Syst. Rev. **36**(SI), 131–146 (2002)
15. Malhotra, B., Nikolaidis, I., Nascimento, M.A.: Aggregation convergecast scheduling in wireless sensor networks. Wirel. Netw. **17**(2), 319–335 (2011)
16. Onus, M., Richa, A., Kothapalli, K., Scheidler, C.: Efficient broadcasting and gathering in wireless ad hoc networks. In: Proceedings of ISPAN, pp. 346–351 (2005)
17. Slater, P.J., Cockayne, E.J., Hedetniemi, S.T.: Information dissemination in trees. SIAM J. Comput. **10**(4), 692–701 (1981)
18. Bonifaci, V., Korteweg, P., Marchetti-Spaccamela, A., Stougie, L.: An approximation algorithm for the wireless gathering problem. Oper. Res. Lett. **36**(5), 605–608 (2008)
19. Wan, P.-J., Huang, S.C.-H., Wang, L., Wan, Z., Jia, X.: Minimum-latency aggregation scheduling in multihop wireless networks. In: Proceedings of the 10th ACM International Symposium on Mobile Ad Hoc Networking and Computing, pp. 185–194. ACM (2009)
20. Minji, W., Xu, J.J., Tang, X.X., Lee, W.-C.: Top-k monitoring in wireless sensor networks. IEEE Trans. Knowl. Data Eng. **19**(7), 962–976 (2007)
21. Xu, X., Wang, S., Mao, X., Tang, S., Li, X.Y.: An improved approximation algorithm for data aggregation in multi-hop wireless sensor networks. In: Proceedings of the 2nd ACM International Workshop on Foundations of Wireless Ad Hoc and Sensor Networking and Computing, FOWANC '09, pp. 47–56. ACM, New York (2009)
22. Yu, B., Li, J., Li, Y.: Distributed data aggregation scheduling in wireless sensor networks. In: Proceedings of the 28th Conference on Computer Communications (INFOCOM), pp. 2159–2167. IEEE (2009)

Author Index

Printed in the United States
By Bookmasters